*Figuring*

*SHAKUNTALA DEVI*

# *Figuring*

*THE JOY OF NUMBERS*

HARPER & ROW, PUBLISHERS
*New York, Hagerstown, San Francisco, London*

FIRST U.S. EDITION

ISBN: 0-06-011069-4

LIBRARY OF CONGRESS CATALOG CARD NUMBER: 78-4731

78 79 80 81 82 10 9 8 7 6 5 4 3 2 1

# *CONTENTS*

Whatever there is in all the three worlds, which are possessed of moving and non-moving beings, cannot exist as apart from the 'Ganita' (calculation)

Mahavira (AD 850)

# *INTRODUCTION*

At three I fell in love with numbers. It was sheer ecstasy for me to do sums and get the right answers. Numbers were toys with which I could play. In them I found emotional security; two plus two always made, and would always make, four – no matter how the world changed.

My interest grew with age. I took immense delight in working out huge problems mentally – sometimes even faster than electronic calculating machines and computers.

I travelled round the world giving demonstrations of my talents. In every country I performed for students, professors, teachers, bankers, accountants, and even laymen who knew very little, or nothing at all, about mathematics. These performances were a great success and everywhere I was offered encouragement and appreciation.

All along I had cherished a desire to show those who think mathematics boring and dull just how beautiful it can be. This book is what took shape.

It is written in the knowledge that there is a range and richness to numbers: they can come alive, cease to be symbols written on a blackboard, and lead the reader into a world of intellectual adventure where calculations are thrilling, not tedious.

To understand the methods I describe you need nothing more than a basic knowledge of arithmetic. There are no formulae, technical terms, algebra, geometry, or logarithms here – if you can carry out the basic functions of adding, subtracting, multiplying, and dividing you can follow all of it. At first glance some of the steps may seem complicated but if you take them slowly you should be able to acquire the skill to solve a wide variety of problems without mistakes – and to do this mentally, or with a minimum use of pencil and paper. But above all I would wish this book to give people an appreciation of the enchantment of numbers.

# 1

# *SOME TERMS DEFINED*

In this book we are not going to be concerned with logarithms or the Calculus, not even with algebra, except in the most elementary way. So those who found that these aspects of mathematics gave them a life-long terror of the whole subject can relax. We are going to be dealing with ordinary numbers which should hold no terrors for anyone.

We will be encountering some mathematical terms and conventions, but most of them are in everyday use. I shall be using them because they are the simplest and clearest means of stating mathematical problems or explanations. But they do all have precise meanings and, even if you are confident that you understand them, it would be as well to read this short chapter so that we can both be sure we are talking about the same things.

**A Digit** is an individual numeral; there are ten of them: 1, 2, 3, 4, 5, 6, 7, 8, 9, and 0.

**A Fraction** in everyday language means a part of a whole. In mathematics it normally has a slightly more precise meaning – a part of a whole expressed by putting one figure over another and dividing them by a line, as in $\frac{1}{2}$ or $\frac{1}{4}$. The figure on top is called the *numerator* (it tells you how many parts you've got) and the one below is called the *denominator* (it tells you what proportion of a whole each part is). They can also be written as 1/2 or 1/4 where the oblique stroke divides numerator and denominator.

**Decimals** are simply a different way of expressing the same thing. Instead of the denominator being variable, as in fractions, in decimals it is always ten, or rather tens, because the system can be extended infinitely to hundredths, thousandths, ten-thousandths, etc of a whole. Thus ·1 is the same as 1/10 and ·01 is the same as 1/100 and ·11 is the same as 11/100.

**Percentage** is yet another way of expressing the same thing. This time the denominator is always 100 – so 4% is the same as ·04 or 4/100. Percentages can be combined with decimals or fractions if it is necessary to express parts smaller than one-hundredth. Thus, one-five-hundredth can be expressed either as ·2% or as $\frac{1}{5}$%.

**Multiplication** is of course to multiply one number by another. The number that is multiplied is called the *multiplicand*, the number by which it is to be multiplied is called the *multiplier* and the result is called the *product* of the two numbers. Thus 2 (multiplier) × 4 (multiplicand) = 8 (product). A *multiple* of a number is simply a number which is made up by multiplying the original number a given number of times. Thus, 2, 4, 6, 8, etc are all multiples of 2.

**Division** is the reverse of multiplication, the division of one number by another. The number to be divided is called the *dividend* and the number by which it is to be divided is called the *divisor*. If the divisor does not go into the dividend an exact number of times, then the figure left over is the *remainder*. For example 9 (dividend) $\div$ 2 (divisor) = 4 with 1 left over as a remainder. If there is no remainder then the result and the divisor are *factors* of the dividend. For instance, 2 and 3 are factors of 6. The factors of a number can also be broken down into their smallest parts; thus the factors of 12 are 2, 2 and 3, or as it is usually put, $2 \times 2 \times 3$.

**Addition** and **subtraction** are so basic that all we need note is that in addition the result is known as the *sum* of the numbers that have been added up; and in subtraction the result is known as the *remainder*, ie, what is left over after one number has been taken from another.

**Roots** and **powers.** If a number is multiplied by itself, the product is the *square* of that number – the square of 2 is 4 and two-squared is written $2^2$. If the square is again multiplied by the original number, ie $2^2 \times 2$, then the result is the *cube* of the number, which is written $2^3$. The process can of course be continued *ad infinitum* – and is called raising a number to its fourth or fifth or 6785th *power*. The *root* of a number is the original number which was squared or cubed or raised to the power of something in order to produce that number. For example the square root of 9 is 3 because $3 \times 3 = 9$. The cube root of 27 is also 3 because $3 \times 9 = 27$, and the fourth root of 81 is also 3 because $3 \times 27 = 81$. The way this is written is $\sqrt{9}$ in the case of a square root and $\sqrt[3]{27}$ or $\sqrt[4]{81}$ in the case of cube, fourth or further roots. The sixth root of 64 is 2; so 2 can be written mathematically as $\sqrt[6]{64}$, just as 64 can be written $2^6$.

We will be encountering a few more technical terms as we go along, but these are the basic ones which you will need right through the book. It's important that, however elementary they seem, you fully grasp what they mean. As I said above they do have precise meanings – for example the difference between the product of 4 and 4 ($4 \times 4 = 16$) and the sum of 4 and 4 ($4 + 4 = 8$) is obviously very important; so it is worth just re-checking to make sure that they are all absolutely clear in your head before going any further.

# 2

# *THE DIGITS*

The ten digits, 1, 2, 3, 4, 5, 6, 7, 8, 9 and 0, are the 'alphabet' of arithmetic. Arranged according to the conventions of mathematics and in conjunction with symbols for addition, subtraction, etc, the digits make up 'sentences' – problems, equations and their solutions. Just as the letters of the alphabet have their own peculiarities and rules (i before e except after c, etc), so the digits each have a particular character in arithmetic. Some of their oddities can be put to good use, others are merely interesting or amusing for their own sake.

In this chapter we are going to look at the digits individually and especially at their multiplication tables. The older of us, at least, will remember sitting in school learning by heart 'one eight is eight, two eights are sixteen, three eights are twenty-four, . . .'. I hope to show you that even in these rather dreary tables there is interest and surprise. For, as well as regular steps which we had to learn by rote, each table contains intriguing 'secret steps' which vary with the different digits. The discovery of these secret steps depends on a technique which is going to crop up throughout this book, and one which I would like to explain at the outset. It very simply consists of adding together the digits in any number made up of two or more digits. For example, with the number 232 the sum of the digits is $2 + 3 + 2 = 7$. If the sum of the digits comes to a two-digit number (for example 456, $4 + 5 + 6 = 15$) then normally you repeat the process ($1 + 5 = 6$).

The secret steps in the multiplication tables are made by applying this technique to the products given by the straight steps. For instance in the 4-times table $3 \times 4 = 12$ (straight step); $1 + 2 = 3$ (secret step); or in the 9-times table $11 \times 9 = 99$; $9 + 9 = 18$, $1 + 8 = 9$.

So, with that technique in mind, let us look at the digits one at a time.

## *NUMBER 1*

The first thing about the number 1 is of course that whatever figure you multiply it by, or divide by it, remains unchanged. The first nine places in the 1-times table therefore have no secret steps, or at least they are the same as the normal ones, but after that it becomes more interesting:

$$
\begin{array}{ll}
10 \times 1 = 10 & 1 + 0 = 1 \\
11 \times 1 = 11 & 1 + 1 = 2 \\
12 \times 1 = 12 & 1 + 2 = 3 \\
13 \times 1 = 13 & 1 + 3 = 4 \\
14 \times 1 = 14 & 1 + 4 = 5 \\
15 \times 1 = 15 & 1 + 5 = 6 \\
16 \times 1 = 16 & 1 + 6 = 7
\end{array}
$$

. . . and so on. No matter how far you go you will find that the secret steps always give you the digits from 1 to 9 repeating themselves in sequence. For example:

| *Straight steps* | *Secret steps* |
|---|---|
| $154 \times 1 = 154$ | $1 + 5 + 4 = 10, 1 + 0 = 1$ |
| $155 \times 1 = 155$ | $1 + 5 + 5 = 11, 1 + 1 = 2$ |
| $156 \times 1 = 156$ | $1 + 5 + 6 = 12, 1 + 2 = 3$ |
| $157 \times 1 = 157$ | $1 + 5 + 7 = 13, 1 + 3 = 4$ |
| $158 \times 1 = 158$ | $1 + 5 + 8 = 14, 1 + 4 = 5$ |
| $159 \times 1 = 159$ | $1 + 5 + 9 = 15, 1 + 5 = 6$ |
| $160 \times 1 = 160$ | $1 + 6 + 0 = 7$ |
| $161 \times 1 = 161$ | $1 + 6 + 1 = 8$ |

In this case it is fairly obvious why the secret steps work, but none the less it is a curiosity that may not have occurred to you before.

Another curiosity about the number 1 is its talent for creating palindromes (numbers that read the same backwards as forwards):

$$1 \times 1 = 1$$
$$11 \times 11 = 121$$
$$111 \times 111 = 12321$$
$$1111 \times 1111 = 1234321$$
$$11111 \times 11111 = 123454321$$
$$111111 \times 111111 = 12345654321$$
$$1111111 \times 1111111 = 1234567654321$$
$$11111111 \times 11111111 = 123456787654321$$
$$111111111 \times 111111111 = 12345678987654321$$

At that point it stops, but the same thing works briefly with the number 11:

$$11 \times 11 = 121$$
$$11 \times 11 \times 11 = 1331$$
$$11 \times 11 \times 11 \times 11 = 14641$$

Finally the fact that 1 is, so to speak, immune to multiplication means that whatever power you raise it to it remains unchanged: $1^{564786} = 1$. By the same token, $\sqrt[8]{1} = 1$.

## *NUMBER 2*

The number 2 has one very obvious characteristic, to multiply any number by 2 is the same as adding it to itself. As for the multiplication table and its secret steps:

| | |
|---|---|
| $2 \times 1 = 2$ | $2$ |
| $2 \times 2 = 4$ | $4$ |
| $2 \times 3 = 6$ | $6$ |
| $2 \times 4 = 8$ | $8$ |
| $2 \times 5 = 10$ | $1 + 0 = 1$ |
| $2 \times 6 = 12$ | $1 + 2 = 3$ |
| $2 \times 7 = 14$ | $1 + 4 = 5$ |
| $2 \times 8 = 16$ | $1 + 6 = 7$ |
| $2 \times 9 = 18$ | $1 + 8 = 9$ |
| $2 \times 10 = 20$ | $2 + 0 = 2$ |
| $2 \times 11 = 22$ | $2 + 2 = 4$ |
| $2 \times 12 = 24$ | $2 + 4 = 6$ |
| $2 \times 13 = 26$ | $2 + 6 = 8$ |
| $2 \times 14 = 28$ | $2 + 8 = 1[0]$ |
| $2 \times 15 = 30$ | $3 + 0 = 3$ |
| $2 \times 16 = 32$ | $3 + 2 = 5$ |
| $2 \times 17 = 34$ | $3 + 4 = 7$ |
| $2 \times 18 = 36$ | $3 + 6 = 9$ |

And so on.

As you can check for yourself, the secret steps go on working *ad infinitum*, always giving you the same sequence of the four even digits followed by the five odd ones.

There is an amusing little party trick that can be played with the number 2. The problem is to express all ten digits, in each case using the number 2 five times and no other number. You are allowed to use the symbols for addition, subtraction, multiplication and division and the conventional method of writing fractions. Here is how it is done:

$$2+2-2-2/2=1$$
$$2+2+2-2-2=2$$
$$2+2-2+2/2=3$$
$$2\times 2\times 2-2-2=4$$
$$2+2+2-2/2=5$$
$$2+2+2+2-2=6$$
$$22\div 2-2-2=7$$
$$2\times 2\times 2+2-2=8$$
$$2\times 2\times 2+2/2=9$$
$$2-2/2-2/2=0$$

Finally, here is another oddity associated with 2:

$$\begin{array}{r} 123456789 \\ +\ 123456789 \\ +\ 987654321 \\ +\ 987654321 \\ +\ 2 \\ \hline 2222222222 \end{array}$$

## *NUMBER 3*

The number three has two distinctions. First of all it is the first triangle number – that is a number the units of which can be laid out to form a triangle, like this $_{0}{}^{0}{}_{0}$. Triangle numbers have importance and peculiarities of their own which we shall encounter later on.

Three is also a prime number, a number that cannot be evenly divided except by itself and by 1. 1 and 2 are of course prime numbers, the next after 3 is 5, then 7, 11, 13, 17, 19 and 23; after that they gradually become increasingly rare. There are a couple of strange things about the first few prime numbers, for example:

$$153 = 1^3 + 5^3 + 3^3$$

And 3 and 5 can also both be expressed as the difference between two squares:

$$3 = 2^2 - 1^2$$
$$5 = 3^2 - 2^2$$

The secret steps in the 3-times table are very simple:

| *Straight steps* | *Secret steps* |
|---|---|
| $3 \times 1 = 3$ | 3 |
| $3 \times 2 = 6$ | 6 |
| $3 \times 3 = 9$ | 9 |
| | |
| $3 \times 4 = 12$ | $1 + 2 = 3$ |
| $3 \times 5 = 15$ | $1 + 5 = 6$ |
| $3 \times 6 = 18$ | $1 + 8 = 9$ |

| | |
|---|---|
| 3 × 7 = 21 | 2 + 1 = 3 |
| 3 × 8 = 24 | 2 + 4 = 6 |
| 3 × 9 = 27 | 2 + 7 = 9 |
| 3 × 10 = 30 | 3 + 0 = 3 |
| 3 × 11 = 33 | 3 + 3 = 6 |
| 3 × 12 = 36 | 3 + 6 = 9 |

Again the pattern of the secret steps recurs whatever stage you carry the table up to – try it and check for yourself.

## *NUMBER 4*

With number 4 the secret steps in the multiplication tables become a little more intricate:

| *Straight steps* | *Secret steps* | | | |
|---|---|---|---|---|
| 4 × 1 = 4 | | | | 4 |
| 4 × 2 = 8 | | | 8 | |
| 4 × 3 = 12 | | 1 + 2 = | | 3 |
| 4 × 4 = 16 | | 1 + 6 = | 7 | |
| 4 × 5 = 20 | | 2 + 0 = | | 2 |
| 4 × 6 = 24 | | 2 + 4 = | 6 | |
| 4 × 7 = 28 | 2 + 8 = 10 | 1 + 0 = | | 1 |
| 4 × 8 = 32 | | 3 + 2 = | 5 | |
| 4 × 9 = 36 | | 3 + 6 = | | 9 |
| 4 × 10 = 40 | | 4 + 0 = | 4 | |
| 4 × 11 = 44 | | 4 + 4 = | | 8 |
| 4 × 12 = 48 | 4 + 8 = 12 | 1 + 2 = | 3 | |
| 4 × 13 = 52 | | 5 + 2 = | | 7 |
| 4 × 14 = 56 | 5 + 6 = 11 | 1 + 1 = | 2 | |
| 4 × 15 = 60 | | 6 + 0 = | | 6 |
| 4 × 16 = 64 | 6 + 4 = 10 | 1 + 0 = | 1 | |

At first the sums of the digits look like a jumble of figures, but choose at random any sequence of numbers and multiply them by 4 and you will see the pattern emerge of two interlinked columns of digits in descending order. For example:

| | | | | |
|---|---|---|---|---|
| $2160\times4=8640$ | $8+6+4+0=18$ | | $1+8=$ | 9 |
| $2161\times4=8644$ | $8+6+4+4=22$ | | $2+2=4$ | |
| $2162\times4=8648$ | $8+6+4+8=26$ | | $2+6=$ | 8 |
| $2163\times4=8652$ | $8+6+5+2=21$ | | $2+1=3$ | |
| $2164\times4=8656$ | $8+6+5+6=25$ | | $2+5=$ | 7 |
| $2165\times4=8660$ | $8+6+6+0=20$ | | $2+0=2$ | |
| $2166\times4=8664$ | $8+6+6+4=24$ | | $2+4=$ | 6 |
| $2167\times4=8668$ | $8+6+6+8=28$ | $2+8=10$ | $1+0=1$ | |
| $2168\times4=8672$ | $8+6+7+2=23$ | | $2+3=$ | 5 |

## *NUMBER 5*

Perhaps the most important thing about 5 is that it is half of 10; as we will see, this fact is a key to many shortcuts in calculation. In the meantime the secret steps in the 5-times table are very similar to those in the 4-times, the sequences simply go upwards instead of downwards:

| *Straight steps* | *Secret steps* | |
|---|---|---|
| $5\times1=5$ | | 5 |
| $5\times2=10$ | $1+0=1$ | |
| $5\times3=15$ | $1+5=$ | 6 |
| $5\times4=20$ | $2+0=2$ | |
| $5\times5=25$ | $2+5=$ | 7 |
| $5\times6=30$ | $3+0=3$ | |
| $5\times7=35$ | $3+5=$ | 8 |

| | | | |
|---|---|---|---|
| $5 \times 8 = 40$ | | $4 + 0 = 4$ | |
| $5 \times 9 = 45$ | | $4 + 5 =$ | 9 |
| $5 \times 10 = 50$ | | $5 + 0 = 5$ | |
| $5 \times 11 = 55$ | $5 + 5 = 10$ | $1 + 0 =$ | 1 |
| $5 \times 12 = 60$ | | $6 + 0 = 6$ | |
| $5 \times 13 = 65$ | $6 + 5 = 11$ | $1 + 1 =$ | 2 |
| $5 \times 14 = 70$ | | $7 + 0 = 7$ | |
| $5 \times 15 = 75$ | $7 + 5 = 12$ | $1 + 2 =$ | 3 |
| $5 \times 16 = 80$ | | $8 + 0 = 8$ | |
| $5 \times 17 = 85$ | $8 + 5 = 13$ | $1 + 3 =$ | 4 |
| $5 \times 18 = 90$ | | $9 + 0 = 9$ | |
| $5 \times 19 = 95$ | $9 + 5 = 14$ | $1 + 4 =$ | 5 |

And so on.

## *NUMBER 6*

This is the second triangle number; and the first perfect number – a perfect number is one which is equal to the sum of all its divisors. Thus, $1 + 2 + 3 = 6$.

The secret steps in the 6-times table are very similar to those in the 3-times, only the order is slightly different.

| *Straight steps* | *Secret steps* | |
|---|---|---|
| $6 \times 1 = 6$ | | 6 |
| $6 \times 2 = 12$ | | $1 + 2 = 3$ |
| $6 \times 3 = 18$ | | $1 + 8 = 9$ |
| $6 \times 4 = 24$ | | $2 + 4 - 6$ |
| $6 \times 5 = 30$ | | $3 + 0 = 3$ |
| $6 \times 6 = 36$ | | $3 + 6 = 9$ |
| $6 \times 7 = 42$ | | $4 + 2 = 6$ |
| $6 \times 8 = 48$ | $4 + 8 = 12$ | $1 + 2 = 3$ |

| | | |
|---|---|---|
| $6 \times 9 = 54$ | | $5 + 4 = 9$ |
| $6 \times 10 = 60$ | | $6 + 0 = 6$ |
| $6 \times 11 = 66$ | $6 + 6 = 12$ | $1 + 2 = 3$ |
| $6 \times 12 = 72$ | | $7 + 2 = 9$ |

And so on.

## *NUMBER 7*

This is the next prime number after 5. The secret steps in the 7-times table almost duplicate those in the 2-times, except that they go up instead of down at each step.

| *Straight steps* | *Secret steps* | |
|---|---|---|
| $7 \times 1 = 7$ | | $7$ |
| $7 \times 2 = 14$ | | $1 + 4 = 5$ |
| $7 \times 3 = 21$ | | $2 + 1 = 3$ |
| $7 \times 4 = 28$ | $2 + 8 = 10$ | $1 + 0 = 1$ |
| $7 \times 5 = 35$ | | $3 + 5 = 8$ |
| $7 \times 6 = 42$ | | $4 + 2 = 6$ |
| $7 \times 7 = 49$ | $4 + 9 = 13$ | $1 + 3 = 4$ |
| $7 \times 8 = 56$ | $5 + 6 = 11$ | $1 + 1 = 2$ |
| $7 \times 9 = 63$ | | $6 + 3 = 9$ |
| $7 \times 10 = 70$ | | $7 + 0 = 7$ |
| $7 \times 11 = 77$ | $7 + 7 = 14$ | $1 + 4 = 5$ |
| $7 \times 12 = 84$ | $8 + 4 = 12$ | $1 + 2 = 3$ |
| $7 \times 13 = 91$ | $9 + 1 = 10$ | $1 + 0 = 1$ |
| $7 \times 14 = 98$ | $9 + 8 = 17$ | $1 + 7 = 8$ |
| $7 \times 15 = 105$ | | $1 + 0 + 5 = 6$ |
| $7 \times 16 = 112$ | | $1 + 1 + 2 = 4$ |
| $7 \times 17 = 119$ | $1 + 1 + 9 = 11$ | $1 + 1 = 2$ |
| $7 \times 18 = 126$ | | $1 + 2 + 6 = 9$ |

$7 \times 19 = 133$ $\qquad$ $1 + 3 + 3 = 7$

$7 \times 20 = 140$ $\qquad$ $1 + 4 + 0 = 5$

And so on.

There is a curious relationship between 7 and the number 142857. Watch:

$$
\begin{aligned}
7 \times 2 &= 7 \times 2 &&= 14 \\
7 \times 2^2 &= 7 \times 4 &&= \quad 28 \\
7 \times 2^3 &= 7 \times 8 &&= \quad\quad 56 \\
7 \times 2^4 &= 7 \times 16 &&= \quad\quad\quad 112 \\
7 \times 2^5 &= 7 \times 32 &&= \quad\quad\quad\quad 224 \\
7 \times 2^6 &= 7 \times 64 &&= \quad\quad\quad\quad\quad 448 \\
7 \times 2^7 &= 7 \times 128 &&= \quad\quad\quad\quad\quad\quad 896 \\
7 \times 2^8 &= 7 \times 256 &&= \quad\quad\quad\quad\quad\quad\quad 1792 \\
7 \times 2^9 &= 7 \times 512 &&= \quad\quad\quad\quad\quad\quad\quad\quad 3584 \\
& && \overline{142857142857142(784)}
\end{aligned}
$$

However far you take the calculation, the sequence 142857 will repeat itself, though the final digits on the right-hand side which I have bracketed will be 'wrong' because they would be affected by the next stage in the addition if you took the calculation on further.

This number 142857 has itself some strange properties; multiply it by any number between 1 and 6 and see what happens:

$$
\begin{aligned}
142857 \times 1 &= 142857 \\
142857 \times 2 &= 285714 \\
142857 \times 3 &= 428571 \\
142857 \times 4 &= 571428 \\
142857 \times 5 &= 714285 \\
142857 \times 6 &= 857142
\end{aligned}
$$

The same digits recur in each answer, and if the products are each written in the form of a circle, you will see that the order of the digits remains the same. If you then go on to multiply the same number by 7, the answer is 999999. We will come back to some further characteristics of this number in Chapter 14.

## *NUMBER 8*

This time the secret steps in the multiplication table are the reverse of those in the 1-times table:

| *Straight steps* | *Secret steps* | |
|---|---|---|
| $8 \times 1 = 8$ | | $8$ |
| $8 \times 2 = 16$ | | $1 + 6 = 7$ |
| $8 \times 3 = 24$ | | $2 + 4 = 6$ |
| $8 \times 4 = 32$ | | $3 + 2 = 5$ |
| $8 \times 5 = 40$ | | $4 + 0 = 4$ |
| $8 \times 6 = 48$ | $4 + 8 = 12$ | $1 + 2 = 3$ |
| $8 \times 7 = 56$ | $5 + 6 = 11$ | $1 + 1 = 2$ |
| $8 \times 8 = 64$ | $6 + 4 = 10$ | $1 + 0 = 1$ |
| $8 \times 9 = 72$ | | $7 + 2 = 9$ |
| $8 \times 10 = 80$ | | $8 + 0 = 8$ |
| $8 \times 11 = 88$ | $8 + 8 = 16$ | $1 + 6 = 7$ |
| $8 \times 12 = 96$ | $9 + 6 = 15$ | $1 + 5 = 6$ |
| $8 \times 13 = 104$ | | $1 + 0 + 4 = 5$ |
| $8 \times 14 = 112$ | | $1 + 1 + 2 = 4$ |
| $8 \times 15 = 120$ | | $1 + 2 + 0 = 3$ |
| $8 \times 16 = 128$ | $1 + 2 + 8 = 11$ | $1 + 1 = 2$ |
| $8 \times 17 = 136$ | $1 + 3 + 6 = 10$ | $1 + 0 = 1$ |

So it continues.

If this is unexpected, then look at some other peculiarities of the number 8:

$$\begin{array}{r} 888 \\ 88 \\ 8 \\ 8 \\ 8 \\ \hline 1000 \end{array}$$

and:

$$\begin{aligned}
88 &= 9 \times 9 + 7 \\
888 &= 98 \times 9 + 6 \\
8888 &= 987 \times 9 + 5 \\
88888 &= 9876 \times 9 + 4 \\
888888 &= 98765 \times 9 + 3 \\
8888888 &= 987654 \times 9 + 2 \\
88888888 &= 9876543 \times 9 + 1
\end{aligned}$$

and, lastly:

$$12345679 \times 8 = 98765432$$

## *NUMBER 9*

With 9 we come to the most intriguing of the digits, indeed of all numbers. First look at the steps in the multiplication table:

| *Straight steps* | *Secret steps* | |
|---|---|---|
| $9 \times 1 = 9$ | | $9$ |
| $9 \times 2 = 18$ | | $1 + 8 = 9$ |
| $9 \times 3 = 27$ | | $2 + 7 = 9$ |
| $9 \times 4 = 36$ | | $3 + 6 = 9$ |
| $9 \times 5 = 45$ | | $4 + 5 = 9$ |
| $9 \times 6 = 54$ | | $5 + 4 = 9$ |
| $9 \times 7 = 63$ | | $6 + 3 = 9$ |
| $9 \times 8 = 72$ | | $7 + 2 = 9$ |
| $9 \times 9 = 81$ | | $8 + 1 = 9$ |
| $9 \times 10 = 90$ | | $9 + 0 = 9$ |
| $9 \times 11 = 99$ | $9 + 9 = 18$ | $1 + 8 = 9$ |
| $9 \times 12 = 108$ | | $1 + 0 + 8 = 9$ |

It is an absolute rule that whatever number you multiply by 9, the sum of the digits in the product will always be 9. Moreover, not only are there no steps in the hidden part of the 9-times table but, for its first ten places, it has another feature:

| | |
|---|---|
| $1 \times 9 = 09$ | $90 = 9 \times 10$ |
| $2 \times 9 = 18$ | $81 = 9 \times 9$ |
| $3 \times 9 = 27$ | $72 = 9 \times 8$ |
| $4 \times 9 = 36$ | $63 = 9 \times 7$ |
| $5 \times 9 = 45$ | $54 = 9 \times 6$ |

The product in the second half of the table is the reverse of that in the first half.

Now, take any number. Say, 87594. Reverse the order of the digits, which gives you 49578. Subtract the lesser from the greater:

$$\begin{array}{r} 87594 \\ -49578 \\ \hline 38016 \end{array}$$

Now add up the sum of the digits in the remainder: $3 + 8 + 0 + 1 + 6 = 18$, $1 + 8 = 9$. The answer will always be 9.

Again, take any number. Say, 64783. Calculate the sum of its digits: $6 + 4 + 7 + 8 + 3 = 28$ (you can stop there or go on as usual to calculate $2 + 8 = 10$, and again, only if you wish, $1 + 0 = 1$).

Now take the sum of the digits away from the original number, and add up the sum of the digits of the remainder. Wherever you choose to stop, and whatever the number you originally select, the answer will be 9.

$$\begin{array}{r} 64783 \\ -28 \\ \hline 64755 \end{array} \qquad 6 + 4 + 7 + 5 + 5 = 27 \qquad 2 + 7 = 9$$

$$\begin{array}{r} 64783 \\ -10 \\ \hline 64773 \end{array} \qquad 6 + 4 + 7 + 7 + 3 = 27 \qquad 2 + 7 = 9$$

$$\begin{array}{r} 64783 \\ -1 \\ \hline 64782 \end{array} \qquad 6 + 4 + 7 + 8 + 2 = 27 \qquad 2 + 7 = 9$$

Take the nine digits in order and remove the 8: 12345679. Then multiply by 9:

$$12345679 \times 9 = 111111111$$

Now try multiplying by the multiples of 9:

$$12345679 \times 18 = 222222222$$
$$12345679 \times 27 = 333333333$$
$$12345679 \times 36 = 444444444$$
$$12345679 \times 45 = 555555555$$
$$12345679 \times 54 = 666666666$$
$$12345679 \times 63 = 777777777$$
$$12345679 \times 72 = 888888888$$
$$12345679 \times 81 = 999999999$$

There is a little mathematical riddle that can be played using 9. What is the largest number that can be written using three digits?

The answer is $9^{9^9}$, or 9 raised to the ninth power of 9. The ninth power of 9 is 387420489. No one knows the precise number that is represented by $9^{387420489}$; but it begins 428124773 . . . and ends . . . 89. The complete number will contain 369 million digits, would occupy over five hundred miles of paper and take years to read.

## *ZERO*

I have a particular affection for zero because it was some of my countrymen who first gave it the status of a number. Though the symbol for a void or nothingness is thought to have been invented by the Babylonians, it was Hindu mathematicians who first conceived of 0 as a number, the next in the progression 4–3–2–1.

Now, of course, the zero is a central part of our mathematics, the key to our decimal system of counting. And it signifies something very different from simply 'nothing' – just think of the enormous difference between ·001, ·01, ·1, 1, 10, and 100 to remind yourself of the importance of the presence and position of a 0 in a number.

The other power of 0 is its ability to destroy another number – zero times anything is zero.

# 3

# *MULTIPLICATION*

Now that we have defined the principal terms we are going to use and have had a look at the digits individually, we can move on to look at some actual calculations and the ways in which they can be simplified and speeded up. I do want to stress at this point that I am not suggesting any magical method by which you can do everything in your head. Do not be ashamed of using paper and pencil, especially when you are learning and practising the different methods I am going to explain. At first you may find you need to scribble down the figures at each intermediate stage in the calculation; with practice you will probably find that it becomes less necessary. But the point is not to be able to do without paper; it is to grasp the methods and understand how they work.

Everyone can do simple multiplication in their head. All mothers are familiar with questions like 4 sausages each for 5 children, or the weekly cost of 2 pints of milk a day at 9 pennies a pint. Problems usually arise when both the figures get into two or three digits – and this is, in many cases, because of the rigid methods we were taught at school. If for example you multiply 456 by 76 by the method usually taught in schools, you end up with a calculation that looks like this:

```
  456
   76
 ----
 2736
31920
 ----
34656
```

You have had to do two separate operations of multiplication and one of addition, and you have had to remember to 'carry' numbers from one column to the next. Many people simply remember the method and do not think about how or why it works.

The first essential thing about multiplication is not to automatically adopt any one method; but to look at the figures involved and decide which of the several methods I am going to explain will be quickest and work best.

## *METHOD I*

The key to this method is the one shortcut in arithmetic that I imagine we all know: that to multiply any number by 10 you simply add a zero; to multiply by 100, add two zeros; by 1000, three zeros; and so on. This basic technique can be widely extended.

To take a simple example, if you were asked to multiply 36 by 5 you would, if you simply followed the method you learned at school, write down both numbers, multiply 6 by 5, put down the 0 and carry the 3, multiply 3 by 5 and add the 3 you carried, getting the correct result of 180. But a moment's thought will show you that 5 is half of 10, so if you multiply by 10, by the simple expedient of adding 0, and then divide by 2 you will get the answer much quicker. Alternatively, you can first halve 36, giving you 18, and then add the 0 to get 180.

Here are some extensions of this method.

To multiply by 15 remember that 15 is one and a half times 10. So to multiply 48 by 15 first multiply by 10

$$10 \times 48 = 480$$

Then, to multiply by 5, simply halve that figure

$$480 \div 2 = 240$$

Add the two products together to get the answer

$$480 + 240 = 720$$

To multiply by $7\frac{1}{2}$ all you have to remember is that $7\frac{1}{2}$ is threequarters of 10. For example $64 \times 7\frac{1}{2}$:

$$64 \times 10 = 640$$

The easiest way of finding three-quarters of 640 is to divide by 4 and multiply by 3, thus:

$$640 \div 4 = 160$$
$$160 \times 3 = 480$$

It's easy to see that this method can work just as well for 75 or 750, and there is no difficulty if the multiplicand is a decimal figure. For example to take a problem in decimal currency, suppose you are asked to multiply 87·60 by 75. Instead of adding a zero just move the decimal point

$$87{\cdot}60 \times 100 = 8760$$
$$8760 \div 4 = 2190$$
$$2190 \times 3 = 6570$$

To multiply by 9 just remember that 9 is one less than 10, so all that is necessary is to add a zero and then subtract the original multiplicand. Take $9 \times 84$.

$$10 \times 84 = 840$$
$$840 - 84 = 756$$

This can be extended; if asked to multiply by 18 all that is necessary is to multiply by 9 and double the product. For example, $448 \times 18$:

$$448 \times 10 = 4480$$
$$4480 - 448 = 4032$$
$$4032 \times 2 = 8064$$

Alternatively you can start from the fact that 18 is 20 less 2 in which case the sequence is:

$$448 \times 2 = 896$$
$$896 \times 10 = 8960$$
$$8960 - 896 = 8064$$

This method can be used for all numbers which are multiples of 9. For example, if asked to multiply 765 by 54 you will realize that 54 is equivalent to $6 \times 9$, the calculation then goes:

$$765 \times 6 = 4590$$
$$4590 \times 10 = 45900$$
$$45900 - 4590 = 41310$$

Multiplying by 11 is easily done if you remember that 11 is $10 + 1$. Therefore to multiply any number by 11 all that is necessary is to add a 0 and then add on the original number. For example, to multiply 5342 by 11:

$$5342 \times 10 = 53420$$
$$53420 + 5342 = 58762$$

In the case of 11 there is an even shorter method that can be used if the multiplicand is a three-digit number in which the sum of the last two digits is not more than 9. For example if asked to multiply 653 by 11, you check that 5 and 3 total less than 9, and since they do, you proceed as follows:

First multiply the last two digits, 53, by 11

$$53 \times 11 = 583$$

To this add the first, hundreds, digit multiplied by 11

$$600 \times 11 = 6600$$
$$6600 + 583 = 7183$$

Another case where this method can be used is with $12\frac{1}{2}$. Here you have a choice, you can either work on the basis that $12\frac{1}{2}$ is 10 plus a quarter of 10, in which case $872 \times 12\frac{1}{2}$:

$$872 \times 10 = 8720$$
$$8720 \div 4 = 2180$$
$$8720 + 2180 = 10900$$

Alternatively, you can work from the basis that $12\frac{1}{2}$ is one-eighth of 100, in which case $872 \times 12\frac{1}{2}$:

$$872 \times 100 = 87200$$
$$87200 \div 8 = 10900$$

It's best to check if the multiplicand is divisible by 8 before adopting the second way. In fact this process of checking pretty well does the calculation for you. If you have to multiply 168 by 12½, a moment's thought shows that 8 goes into 168 exactly 21 times. All you then have to do is add two zeros to get the correct product – 2100.

But if the multiplicand had been, say, 146, then clearly the first method is the one to use.

Here are some other relationships that can be exploited to use this basic method:

112½ is 100 plus one-eighth of 100

125 is 100 plus one-quarter of 100; or 125 is one-eighth of 1000

45 is 50 minus 5, 50 is half of 100 and 5 is one-tenth of 50

25 is a quarter of 100

35 is 25 plus 10

99 is 100 minus 1

90 is 100 minus one-tenth of 100

and so on. If you experiment you will find many more of these useful relationships, all of which can be used to take advantage of the basic shortcut offered by the fact that to multiply by 10 all you do is add a zero. After some practice you will find that you can spot almost without thinking a case where this first method is going to help.

## *METHOD II*

The next method I am going to describe can be used when the multiplier is a relatively small number, but one for which our first method is unsuitable because there is no simple relationship to 10 which can be spotted and exploited.

We have in fact already made use of this second method without my drawing attention to it. When, for instance, you multiply by 20 by the simple and obvious means of multiplying by 2 and then adding a 0, what you are doing is taking out the factors of 20, 2 and 10, and multiplying by them each in turn. This method can be extended to any number which can be broken down into factors. For example if you are asked to multiply a number by 32 you can break 32 down into its factors, 8 and 4, and proceed as follows (here, to start with anyway, I suggest the use of paper and pencil):

To multiply 928 by 32

```
  928
    4
 ----
 3712
    8
 ----
29696
```

Even if you have to write the calculation out as above it is a great deal quicker than the conventional method, which would not only involve two sequences of multiplication but also one of addition.

In practice this method is best used where the multiplier is relatively small and its factors therefore easily and quickly extracted; if you remember the multiplication tables up to 12 you will be able to judge at a glance whether or not this is a suitable method in the case of a two-digit multiplier.

## *METHOD III*

This third method is one that should be used where there is no easily discernible means of using the first two, and where the number of digits involved makes them impracticable.

I am going to start by explaining the method with relatively small numbers so that you can grasp the essentials.

For our first example let us take $13 \times 19$. First, add the unit digit of one number to the other number, thus:

$$9 + 13 = 22$$

You then think of that sum as so many tens, in this case 22 tens. Now multiply the two unit digits together:

$$3 \times 9 = 27$$

Finally add the product to the tens figure you already have:

$$27 + 22 \text{ (tens)} = 247$$

Here is another example, $17 \times 14$:

$$4 + 17 = 21 \text{ (tens)}$$
$$7 \times 4 = 28$$
$$28 + 21 \text{ (tens)} = 238$$

There are particular short cuts for multiplying together two-figure numbers with the same tens or the same units figure.

When the tens figure is common, you add to one number the units figure of the other, multiply this sum by the common tens figure and add, to this product, considered as tens, the product of the two units digits. For example, $49 \times 42$:

$$49 + 2 = 51$$
$$51 \times 4 = 204$$

Add to this product (2040, when considered as tens) the product of the units ($9 \times 2 = 18$) to get the final product of 2058.

Another example, $58 \times 53$:

$$58 + 3 = 61$$
$$61 \times 5 = 305 \text{ (tens)}$$
$$8 \times 3 = 24$$
$$3050 + 24 = 3074$$

If the common tens figure is 9 there is an even simpler method. Subtract each of the numbers from 100. Multiply the remainders together – this gives you the last two figures of the final product. To arrive at the first two digits take away from one of the numbers the figure by which the other was short of 100. For example, $93 \times 96$:

$$100 - 93 = 7$$
$$100 - 96 = 4$$

$4 \times 7 = 28$, your two last digits. $93 - 4$ or $96 - 7$ gives you 89, your first two digits. The product is therefore 8928.

If the units figures of two two-figure numbers are the same you can multiply them by adding the product of the two tens figures (considered as hundreds) to the sum of the tens figures multiplied by the common units digit, and the square of the common units digit. For example, $96 \times 46$:

$$9 \times 4 = 36 \text{ (hundreds)}$$
$$9 + 4 = 13,\ 13 \times 6 = 78 \text{ (tens)}$$
$$6 \times 6 = 36$$
$$3600 + 780 + 36 = 4416$$

Or, to multiply 62 by 42:

$$6 \times 4 = 24 \text{ (hundreds)}$$
$$6 + 4 = 10,\ 10 \times 2 = 20 \text{ (tens)}$$
$$2 \times 2 = 4$$
$$2400 + 200 + 4 = 2604$$

If the common final digit is 5 it is even simpler – to the product of the two tens figures considered as hundreds add half the sum of the two tens figures, still considered as hundreds. Then add 25 to arrive at the final product. For example, $45 \times 85$:

$$4 \times 8 = 32 \text{ (hundreds)}$$
$$\frac{4+8}{2} = 6 \text{ (hundreds)}$$
$$3200 + 600 = 3800$$

Add 25 to obtain the final product of 3825.

If the sum of the tens is an odd number take it to the nearest whole number below, and add 75, not 25, to obtain the final product. For example, $95 \times 25$:

$$9 \times 2 = 18 \text{ (hundreds)}$$
$$\frac{9+2}{2} = 5{\cdot}5 \text{ (hundreds)}$$
$$1800 + 500 + 75 \text{ (25 plus ·5 of 100)}$$
$$= 2375$$

All of the methods I have described so far can be done mentally when you have had a little practise – I will now describe others which can be used more generally, but which require pencil and paper. Even with these methods most of the calculations can be done mentally; you use the paper to keep note of your intermediate results. In each case you do a sequence of diagonal or vertical multiplications – the pattern of these is shown in diagrams to the right of the examples.

For example, to multiply 63 by 48 write down the numbers thus:

```
63
48
—
```

$3 \times 8 = 24$, so put 4 in the units column and carry the 2; you will add this to the sum of the products of the 'diagonals' – 8 and 6, and 4 and 3. Your mental calculation runs thus:

$$6 \times 8 = 48; 4 \times 3 = 12; 48 + 12 + 2 = 62.$$

You write 2 in the tens column and carry 6 – this you add to the product of the two tens digits – 6 and 4.

The mental calculation $6 \times 4 = 24$; $24 + 6 = 30$ gives the final figures – all you have had to write down is the problem itself and the answer:

```
  63
  48
————
3024
```

Set out three-digit figures to be multiplied together in the same way:

```
436
254
———
```

Again multiply the units digits and write down the units figure of the answer, 4, and carry the tens digit, 2. Now multiply 3 by 4 and add the 2 you are carrying to make 14. Add this to the product of $5 \times 6 = 30$, to make a total of 44. You now have two figures of your final answer and are still carrying only one figure – 4 – in your head. The figures you have written read

$$\begin{array}{r} 436 \\ 254 \\ \hline 44 \end{array}$$

Your next mental steps are to add the 4 you are carrying to the products of $4 \times 4$, $6 \times 2$, and $3 \times 5$, and the calculation will run: $4 \times 4 = 16$; $16 + 4 = 20$; $6 \times 2 = 12$; $12 + 20 = 32$; $3 \times 5 = 15$; $15 + 32 = 47$.

Set down the 7 and carry the 4.

Now multiply the left-hand set of diagonals – 4 and 5, and 3 and 2 and add the carried 4: $4 \times 5 = 20$; $20 + 4 = 24$; $3 \times 2 = 6$; $6 + 24 = 30$.

Set down the 0, carry the 3, and add it to the product of the first hundreds digits, 4 and 2: $4 \times 2 = 8$; $8 + 3 = 11$.

Now write 11 next to the 0:

$$\begin{array}{r} 436 \\ 254 \\ \hline 110744 \end{array}$$

Again, all you have had to write are problem and answer.

If the multiplier has only two figures you can still use this method by replacing the missing hundreds figure with a 0. For example set out $476 \times 26$ like this:

476
026
——

$6 \times 6 = 36$; put down the 6, carry 3, and proceed as before.

$7 \times 6 = 42$; plus 3 carried $= 45$; plus $2 \times 6\,(= 12) = 57$. Put down the 7 and carry the 5.

$4 \times 6 = 24$; plus the 5 carried $= 29$; $6 \times 0 = 0$, $2 \times 7 = 14$. $29 + 14 = 43$. Set down 3 and carry 4.

$4 \times 2 = 8$; $8 + 4 = 12$; $7 \times 0 = 0$; $12 + 0 = 12$ – so you write 2 and carry 1. The figures you have now written read:

476
026
——
2376

$4 \times 0 = 0$, but you are still carrying 1 so the final answer reads 12376.

We can now extend the method to deal with four-figure numbers. For instance to multiply 9246 by 2541 set down the problem in the same way as before:

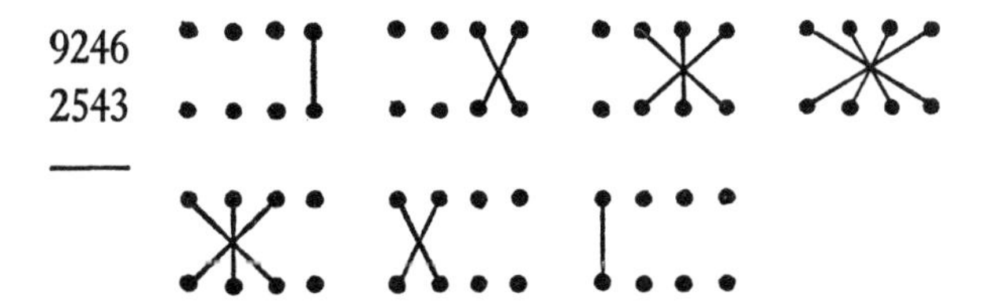


First the units: $3 \times 6 = 18$, write down the 8 carry the 1. Next the first pair of diagonals: $4 \times 3 = 12$; $12 + 1 = 13$; $4 \times 6 = 24$; $24 + 13 = 37$.

Write down the 7 and carry the 3:

$$\begin{array}{r} 9246 \\ 2543 \\ \hline 78 \end{array}$$

Now $2 \times 3$, $4 \times 4$, and $5 \times 6$:

$$\begin{array}{lr} 2 \times 3 = \ \ 6 & 6 + 3 \text{ (carried)} = \ \ 9 \\ 5 \times 6 = 30 & 30 + 9 = 39 \\ 4 \times 4 = 16 & 16 + 39 = 55 \end{array}$$

Set down 5 and carry 5:

$$\begin{array}{r} 9246 \\ 2543 \\ \hline 578 \end{array}$$

Now the four sets of diagonals:

$$\begin{array}{lr} 9 \times 3 = 27 & 27 + 5 \text{ (carried)} = 32 \\ 2 \times 6 = 12 & 12 + 32 = 44 \\ 2 \times 4 = 8 & 8 + 44 = 52 \\ 5 \times 4 = 20 & 20 + 52 = 72 \end{array}$$

Set down 2 and carry 7:

$$\begin{array}{r} 9246 \\ 2543 \\ \hline 2578 \end{array}$$

By the same procedures as before:

$$9 \times 4 = 36 \qquad 36 + 7 \text{ (carried)} = 43$$
$$4 \times 2 = 8 \qquad 43 + 8 = 51$$
$$2 \times 5 = 10 \qquad 51 + 10 = 61$$

gives us another digit of the solution, and 6 to carry.

$$9 \times 5 = 45 \qquad 45 + 6 \text{ (carried)} = 51$$
$$2 \times 2 = 4 \qquad 51 + 4 = 55$$

gives another, 5, with 5 to carry:

```
  9246
  2543
------
512578
```

and we get the final two digits by multiplying 9 × 2 and adding the 5 we are carrying: 18 + 5 = 23

```
    9246
    2543
--------
23512578
```

This same method of adding the products of groups of diagonal multiplications can be used for larger numbers, but a little more paperwork will make it simpler. For instance to multiply 637432 by 513124 start by setting the numbers out in the usual way:

```
637432
513124
------
```

then proceed as follows:

$(4 \times 2) =$ 8

$(4 \times 3) + (2 \times 2) = 12 + 4 =$ 16

$(4 \times 4) + (2 \times 1) + (3 \times 2) = 16 + 2 + 6 =$ 24

$(4 \times 7) + (2 \times 3) + (3 \times 1) + (4 \times 2) =$
$28 + 6 + 3 + 8 =$ 45

$(4 \times 3) + (2 \times 1) + (2 \times 7) + (3 \times 3) + (4 \times 1) =$
$12 + 2 + 14 + 9 + 4 =$ 41

$(4 \times 6) + (5 \times 2) + (2 \times 3) + (1 \times 3) + (7 \times 1) +$
$(3 \times 4) = 24 + 10 + 6 + 3 + 7 + 12 =$ 62

$(2 \times 6) + (5 \times 3) + (1 \times 4) + (3 \times 1) + (7 \times 3) =$
$12 + 15 + 4 + 3 + 21 =$ 55

$(1 \times 6) + (5 \times 4) + (1 \times 7) + (3 \times 3) =$
$6 + 20 + 7 + 9 =$ 42

$(3 \times 6) + (5 \times 7) + (3 \times 1) = 18 + 35 + 3 =$ 56

$(1 \times 6) + (5 \times 3) = 6 + 15 =$ 21

$(6 \times 5) =$ 30

327081657568

All the computing goes on in your head, the addition at the end involves only two digits in each column and the amount of paperwork is limited to the figures on the right-hand side of the page.

## *CHECKING BY CASTING OUT THE NINES*

No one is immune from error, and quick ways of checking long computations are always useful. The method I describe here is not infallible, but if the answer you have obtained checks out as correct when you have used it, the chances of your being wrong are very slim indeed. This is the way it is done:

Add up the digits in each factor (the numbers you are multiplying together) and in the product.

Divide each of the sums of digits by 9.

Set down the remainder in each case – this is known as the check number.

Multiply the check number of the multiplicand by the check number of the multiplier, and add up the digits in the product.

If this gives you the same number as the check number of the original product you can assume that that product was correct.

For example to check that $8216 \times 4215 = 34630440$:

$8 + 2 + 1 + 6 = 17$, casting out the nines leaves 8 as check number.

$4 + 2 + 1 + 5 = 12$, casting out the nines leaves 3 as check number.

$3 + 4 + 6 + 3 + 0 + 4 + 4 + 0 = 24$, casting out the nines leaves 6 as check number.

$3 \times 8 = 24$, $2 + 4 = 6$, which was the check number of the original product.

# 4

# *ADDITION*

Addition can be thought of as an extension of counting. It is involved in almost all calculations – it is simple, but mistakes are easy to make when long lists of numbers are being added together, or when numbers are carried from one column to the next.

One way of avoiding mistakes in long additions – of twenty or thirty numbers say – is to break the list up into smaller groups and then add the totals of all the groups together. Remember, too, that it is easier to add round numbers like 40 or 150 than numbers ending in 7, 8, or 9. If you round these awkward numbers up by adding 3, 2, or 1 the calculation is not much longer, and is easier. Instead of

$$49 + 52 = 101$$

think of it as

$$\begin{aligned} 50 \text{ (that is } 49 + 1) + 52 &= 102 - 1 \\ &= 101 \end{aligned}$$

Carrying errors very often occur when you are adding in your head, and have to keep both the digits of the final answer and those you are carrying to add to the next column in your mind at the same time. One way of avoiding this sort of confusion is to reverse the ordinary method of working and add from left to right instead of from right to left. If the numbers are small no carrying is involved and the final total is arrived at by way of partial totals. For instance to add 353 to 2134 first add 300 (giving the partial total of 2434), then 50 (giving 2484) and then 3 to arrive at the final total of 2487. Or, to add 5128 to 2356, increase 5128 by steps of 2000, 300, 50, and 6 to arrive at the partial totals of 7128, 7428, 7478 and the final total of 7484.

You can also work from left to right when adding up columns of numbers. For example if you are adding

```
215
426
513
112
328
———
```

first add the figures in the hundreds column and hold the total, 1500, in your head. Now add the total of the tens column, 70, to it. To this total of 1570 add the sum of the units column, 24, to arrive at the final total of 1594.

Even when you use pencil and paper carrying errors can occur. Here is a method of working which makes them much less likely:

```
 9256384
 5678256
 8143132
 1829527
 6415948
 ———————
 9191027
2213222
————————
31323247
```

Add each column of figures separately. Put the units figure of the total directly below the column you have just added, the tens figure one line down and one place to the left. Repeat this process for each column and add the two rows of figures together to obtain the final result.

Another method of avoiding carrying also involves adding each column separately. The column totals are set out in a staggered line, the units figure of the second column below the tens figure of the first, the units figure of the third column below the tens figure of the second and so on. The column totals are then added to give the final answer. Here is an example:

```
962853         19
524861        22
212346       24
401258      13
864321     15
          28
          -------
          2965639
```

(You can of course work from left to right if you wish and set each column total one place further right rather than one place further left.)

```
962853    28
524861     15
212346      13
401258       24
864321        22
               19
          -------
          2965639
```

There are even tricks you can play with addition. Ask a friend to write down any five-figure number – say he writes 21564. You, apparently at random, choose figures to write below his. You put down 78435 – these are in fact the digits which, added to those they stand below, will total 9. You now ask your friend to add a further five-figure number, and you again write below his digits those that would make them up to 9. Your friend adds a final fifth line of five figures and you instantly draw a line and add all five numbers together – the total will always be the last number your friend wrote with 2 subtracted from the last digit and 2 inserted before the first one.

```
 21564
 78435
 12564
 87435
 42145
 -----
242143
```

## *CHECKING BY CASTING OUT THE NINES*

Additions, like multiplications, can be checked (though not infallibly checked) by casting out the nines.

First add up the digits of each number in the sum, and of the answer you had arrived at. Divide each of these numbers by 9 and set down the remainder. Total the remainders of the numbers you were adding together, and again cast out the nines. If the remainder you are left with is the same as the remainder when the sum of the digits of your answer was divided by 9 you may assume that answer was correct.

# 5

# *DIVISION*

Division is the reverse of multiplication – just as subtraction is the reverse of addition. To multiply 3 by 4 is to add four threes together, and find the total – 12. Dividing 12 by 3 could be said to be subtracting threes until you had nothing left – and you would of course do it four times. But division seems more intractable to most people, and particularly difficult to handle mentally. In this chapter I will show how division by particular numbers can be handled in particular ways, and, at the end of the chapter, how you can check to see whether a number is or is not divisible by any number from 2 to 11 (or, of course, any multiple of such a number). But first some special cases.

**To divide a number by 5** you take advantage of the fact that 5 is half of 10, multiply the dividend (the number being divided) by 2 and divide by 10 by moving the decimal point one place to the left. For example:

$$165 \text{ divided by } 5 = \frac{330}{10} = 33$$

**To divide by 15** multiply the dividend by 2 and divide by 30. For example

$$105 \text{ divided by } 15 = \frac{210}{30} = 7$$

If you simplify in this way – by doubling divisor and dividend – and are left at the end of the calculation with a remainder you must remember that this, too, will be doubled. Some divisions can be simplified by halving both divisor and dividend – divisions by 14, 16, 18, 20, 22, and 24 become divisions by 7, 8, 9, 10, 11, and 12 – but this time you must remember to double the remainder. Here are some examples:

$$392 \text{ divided by } 14 = \frac{196}{7} = 28$$

$$464 \text{ divided by } 16 = \frac{232}{8} = 29$$

$$882 \text{ divided by } 18 = \frac{441}{9} = 49$$

$$4960 \text{ divided by } 20 = \frac{2480}{10} = 248$$

$$946 \text{ divided by } 22 = \frac{473}{11} = 43$$

$$1176 \text{ divided by } 24 = \frac{588}{12} = 49$$

## *DIVIDING BY FACTORS*

If a number can be broken down into factors, it may be simpler to divide by these, successively, than to do a single calculation. A mental division by 8, and then by 4, is simpler than a division by 32. For example to divide 1088 by 32 first divide by 8:

$$\frac{1088}{8} = 136$$

and then divide the answer, 136, by 4 to get the final result – 34.

Numbers easy to handle in this way are the products in the basic multiplication tables – multiples of 11 for instance go particularly smoothly:

To divide 2695 by 55, first divide by 5:

$$\frac{2695}{5} = 539,$$

and then by 11:

$$\frac{539}{11} = 49$$

to get the answer.

To divide by numbers that are powers of 2 (4, 8, 16, and so on), you merely have to go on halving the dividend. 16 for instance is $2^4$ so halving the dividend four times is the same as dividing by 16.

To divide 8192 by 16
halve once to get 4096
and again to get 2048
and a third time to get 1024
and finally a fourth time to get the answer 512

This technique makes dividing by high powers of 2 easy – for instance to divide 32768 by 128 – which is $2^7$:

32768 halved = 16384
16384 halved = 8192
8192 halved = 4096
4096 halved = 2048
2048 halved = 1024
1024 halved = 512
and 512 halved = 256 – which is the required answer.

The methods I have described so far work only with some numbers – thinking about what sort of numbers are involved in a calculation is often the first step to finding a quick way to do it. The more skilled you get at mental multiplication and division the more steps in a normal division sum you will be able to do in your head. For instance in the sum set out below only the remainders are noted down – partial products are arrived at and the subtractions done mentally:

```
31)13113(423
     71
      93
      —
      00
```

## *DIVIDING BY FRACTIONS MENTALLY*

It is easier to divide by whole numbers than fractions – if both divisor and dividend are multiplied by the same factor the answer to the problem will be the same. (Any remainder, however, will be a multiple or fraction of the correct value.) If, for instance, you are dividing by $7\frac{1}{2}$ it is simpler to multiply both numbers in the calculation by 4 – dividing four times your original dividend by 30. For example to divide 360 by $7\frac{1}{2}$:

$$360 \times 4 = 1440$$
$$7\tfrac{1}{2} \times 4 = 30$$
$$\frac{1440}{30} = 48$$

By the same principle when dividing by 12½ multiply the dividend by 8 and divide by 100, when dividing by 37½ multiply by 8 and divide by 300, and when dividing by 62½ multiply by 8 and divide by 500.

To divide by 1½ double the dividend and the divisor and divide by 3 (but remember to halve any remainder).

To divide a number by 2½ double the dividend and the divisor and divide by 5 – and go about dividing by 3½ in the same way.

## *CHECKING WHETHER A NUMBER IS EXACTLY DIVISIBLE*

There are tests that can be made to show whether a number is exactly divisible by another number (or multiple of it). Here are some of them:

If a number is divisible by **two** it will end in an even number or a 0.

If a number is divisible by **three** the sum of its digits will be divisible by 3 (for example $372 = 3 + 2 + 7 = 12$). A corollary of this is that any number made by rearranging the digits of a number divisible by 3 will also be divisible by 3.

If a number is divisible by **four** the last two digits are divisible by 4 (or are zeros).

If a number is divisible by **five** the last digit will be 5 or 0.

If a number is divisible by **six** the last digit will be even and the sum of the digits divisible by 3.

There is no quick test for finding out if a number is divisible by **seven.**

If a number is divisible by **eight** the last three digits are divisible by 8.

If a number is divisible by **nine** the sum of its digits is divisible by 9.

If a number is divisible by **ten** it ends with a 0.

If a number is divisible by **eleven** the difference between the sum of the digits in the even places and the sum of the digits in the odd places is 11 or 0. For example, 58432 is shown to be divisible by 11 because

$$\begin{aligned} 5 + 4 + 2 &= 11 \\ 8 + 3 &= 11 \\ 11 - 11 &= 0 \end{aligned}$$

and 25806 is shown to be divisible by 11 because:

$$\begin{aligned} 2 + 8 + 6 &= 16 \\ 5 + 0 &= 5 \\ 16 - 5 &= 11 \end{aligned}$$

## *CHECKING BY CASTING OUT THE NINES*

Divisions, like additions and multiplications, can be checked by casting out the nines – although, again, the method is not infallible.

You obtain check numbers, as before, by adding up the digits in the numbers in the calculation – in this case the divisor, dividend, and quotient – and dividing them by 9. The check number is the remainder left after each division.

Multiply the check number of the divisor by the check number of the quotient. If the check number of this product is the same as the check number of the dividend the division may be assumed to be correct. For example to check:

$$\frac{2426376}{5321} = 456$$

$2 + 4 + 2 + 6 + 3 + 7 + 6 = 30$, casting out the nines leaves 3
$5 + 3 + 2 + 1 = 11$, casting out the nines leaves 2
$4 + 5 + 6 = 15$, casting out the nines leaves 6

$6 \times 2 = 12$ (check number divisor $\times$ check number quotient). Casting out the nines leaves 3 – which is the same as the check number of the dividend.

When there is a remainder the sum of the digits in the remainder are added to the product of the check numbers of the quotient and the divisor, and the operation is completed in the usual way. For example to check:

$$\frac{1481265}{4281} = 346 \text{ remainder } 39.$$

$1 + 4 + 8 + 1 + 2 + 6 + 5 = 27$, casting out the nines leaves 0
$4 + 2 + 8 + 1 = 15$, casting out the nines leaves 6

$3 + 4 + 6 = 13$, casting out the nines leaves 4
$3 + 9 = 12$, casting out the nines leaves 3
$6 \times 4 = 24$, $24 + 3 = 27$, adding these digits and casting out the nines leaves 0 – which is the check number of the dividend.

# 6

# *SUBTRACTION*

Subtraction is the opposite of addition, but many people find adding easier. If the number being subtracted is a multiple of 10 however the sum is easy, and you can simplify subtractions by adding to both the number being subtracted, and the number being subtracted from the figure which will bring the former up to round figures. It is a particularly useful trick when subtracting sums of money in decimal currencies. For example to subtract 46 from 58 add 4 to both numbers – the problem then becomes the easy one of

$$62 - 50 = 12$$

Or, to deduct $1·72 from $3·64, add 8 cents to each sum:

$$\$3{\cdot}72 - \$1{\cdot}80 = \$1{\cdot}92.$$

When you are dealing with three-figure numbers you will bring the number being subtracted up to the nearest hundred: 246 – 182 becomes 264 – 200 and the answer, 64, obvious.

And with four-figure numbers you add to both numbers the number which will bring the number being subtracted up to the nearest thousand. When 2348 – 1821 becomes 2527 – 2000, there is no problem in seeing that 527 is the answer.

## *CHECKING ERRORS IN SUBTRACTION*

The simplest check is to add your answer to the number being subtracted – the sum will be the same as the number being subtracted from, if you have made no mistakes. You can also use the method of casting out the nines, which I have described already – in this case the procedure is to add up the digits of each number in the subtraction, divide by 9, and note the check numbers – the remainders – arrived at. Having cast out the nines in this way subtract the check number of the number being subtracted, from the check number of the number being subtracted from; if the figure arrived at is the same as the check number of your answer you can assume that your original subtraction was carried out correctly. But again remember that casting out the nines is not infallible.

# 7

# *GCM and LCM*

I have already explained how numbers can be broken down into factors – numbers which, multiplied together, give the original number as a product.

The Greatest Common Measure, or Highest Common Factor, is the greatest number that is a factor common to two numbers – it can be defined as 'the greatest number that is contained an exact number of times in each of two or more numbers' and abbreviated to GCM or HCF.

The usual way of finding the GCM of two numbers is to break both of them down into their factors, pick out the factor or factors they have in common, and multiply these together.

For instance to find the GCM of 72 and 126:

$$72 = 2 \times 2 \times 2 \times 3 \times 3$$
$$126 = 7 \times 2 \times 3 \times 3$$

The common factors are 2, 3, and 3, and the GCM, the product of these factors multiplied together, 18.

Or to find the GCM of 7293 and 442:

$$7293 = 3 \times 11 \times 17 \times 13$$
$$442 = 2 \times 13 \times 17$$

The GCM is therefore $13 \times 17$, or 221.

As larger numbers are harder to split into factors another method can be used. This involves dividing the larger number by the smaller, and then dividing the smaller number by the remainder from the first division. The remainder of the first division is then divided itself by the remainder from the second, and so on, until no remainder is left. The last divisor is the GCM of the two numbers you started with. Here are two examples:

Find the GCM of 4781 and 6147.

```
4781)6147(1
     4781
     ----
     1366)4781(3
          4098
          ----
           683)1366(2
               1366
               ----
```

The required GCM is 683.

Find the GCM of 13536 and 23148.

```
13536)23148(1
      13536
      -----
       9612)13536(1
             9612
             ----
             3924)9612(2
                  7848
                  ----
                  1764)3924(2
                       3528
                       ----
                        396)1764(4
                            1584
                            ----
                             180)396(2
                                 360
                                 ---
                                  36)180(5
                                     180
                                     ---
```

The required GCM is 36.

## *LEAST COMMON MULTIPLE*

The Least Common Multiple is the smallest number which contains an exact number of each of two or more numbers. The usual method of finding the Least Common Multiple (LCM) of two or more numbers is to break them down into their prime factors – factors which cannot themselves be split into smaller factors.

From the prime factors of all the numbers choose the highest powers of each factor that occurs. Multiply these together, and the product will be the required LCM.

For example, to find the LCM of 78, 84, and 90:

$$78 = 2 \times 13 \times 3$$
$$84 = 2 \times 2 \times 7 \times 3 = 2^2 \times 7 \times 3$$
$$90 = 2 \times 3 \times 3 \times 5 = 2 \times 3^2 \times 5$$

Therefore the LCM is $2^2 \times 3^2 \times 13 \times 7 \times 5 = 16380$.

To find the prime factors set it out in this way (the problem is to find the prime factors and LCM of 192, 204, 272):

```
2)192, 204, 272
  ------------
2) 96, 102, 136
  ------------
2) 48,  51,  68
  ------------
2) 24,  51,  34
  ------------
3) 12,  51,  17
  ------------
17) 4,  17,  17
  ------------
    4,   1,   1
```

Each number is divided by factors common to at least two of them. When no more divisions are possible, multiply together all the numbers you have divided by, and the numbers in the bottom row, to obtain the least common multiple.

$$2 \times 2 \times 2 \times 2 \times 3 \times 17 \times 4 = 3264$$

When the numbers involved are very large it will be quicker to use the GCM of the numbers as the first divisor:

Find the LCM of 7535 and 11645.

The GCM of 7535 and 11645 is found to be 685. Dividing by the GCM, we obtain:

$$\begin{array}{r|cc} 685 & 7535, & 11645 \\ \hline & 11 & 17 \end{array}$$

Therefore the LCM is

$$685 \times 11 \times 17 = 128095$$

Find the LCM of 4781 and 6147.

The GCM of 4781 and 6147 is found to be 683.

$$\frac{4781 \times 6147}{683} = 43029$$

The lowest common multiple of two numbers is the same as the product of the two numbers divided by their highest common factor. (Conversely the GCM and the LCM multiplied together gives the product of the two numbers.)

When the LCM of a list of numbers is wanted find the LCM of the first two numbers, then the LCM of the third number and the LCM of the first two, and so on:

Find the LCM of 385, 231, 165, and 105.

The GCM of 385 and 231 is 77. Therefore the LCM of 385 and 231 is

$$\frac{385 \times 231}{77} = 1155$$

The GCM of 165 and 1155 is 165. Therefore the LCM of 165 and 1155 is

$$\frac{165 \times 1155}{165} = 1155$$

Now the GCM of 1155 and 105 is 105. Therefore the LCM of 1155 and 105 is

$$\frac{1155 \times 105}{105} = 1155$$

# 8

# *SQUARES and SQUARE ROOTS*

The square of a number is that number, multiplied by itself. It is also – and this explains why it is called a square – the number of units in a square grid having sides of a given number of units.

Think of building up squares by laying down marbles – the first square will have one: o, the second $\begin{smallmatrix} o & o \\ o & o \end{smallmatrix}$, the third $\begin{smallmatrix} o & o & o \\ o & o & o \\ o & o & o \end{smallmatrix}$ and so on. One, four, and nine, are thus the squares.

So 4 squared is $4 \times 4$. It is written with a figure $^2$ ($4^2$) and is known as the second power of four. Four squared equals 16, and 4 is therefore the square root of 16 – which is written thus: $\sqrt{16}$.

The sequence of squares and the sequence of odd numbers have what at first sight seems a mysterious relationship:

$$1 = 1 = 1^2$$
$$1 + 3 = 4 = 2^2$$
$$1 + 3 + 5 = 9 = 3^2$$
$$1 + 3 + 5 + 7 = 16 = 4^2$$
$$1 + 3 + 5 + 7 + 9 = 25 = 5^2$$
$$1 + 3 + 5 + 7 + 9 + 11 = 36 = 6^2$$
$$1 + 3 + 5 + 7 + 9 + 11 + 13 = 49 = 7^2$$

And so on.

But if you think of the square of marbles the reason for it becomes clear:

1 2 3 4 5 6 7 8 9

To make up the second square one must add 3 marbles to the 1 square. To make up the third square one must add 5, and so on.

This diagram also makes it clear why another relationship holds: any triangle number added to the next highest triangle number always gives a square number. (Triangle numbers, as their name suggests, are those that can be made from units in a triangular display: 3, 6, 10, and so on.)

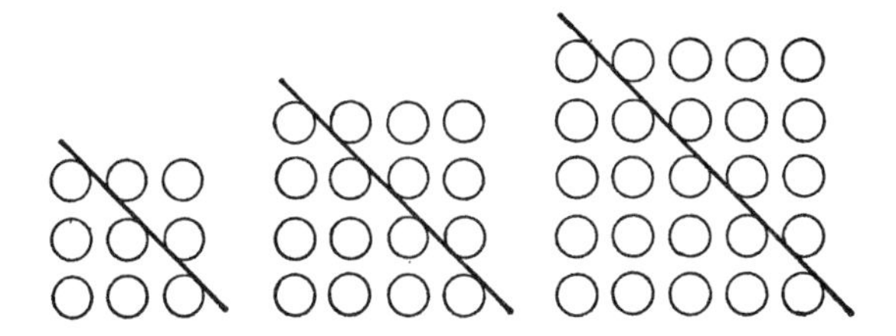

The triangular numbers have another special relationship to the square numbers: multiply any triangle number by 8 and add 1 and you have a square number. The sequence runs:

$$\begin{array}{lll} \text{Triangle } 1 = 1 & (8 \times 1) + 1 = 9 & = 3^2 \\ \text{Triangle } 2 = 3 & (8 \times 3) + 1 = 25 & = 5^2 \\ \text{Triangle } 3 = 6 & (8 \times 6) + 1 = 49 & = 7^2 \\ \text{Triangle } 4 = 10 & (8 \times 10) + 1 = 81 & = 9^2 \\ \text{Triangle } 5 = 15 & (8 \times 15) + 1 = 121 & = 11^2 \end{array}$$

Note that the odd sequence of numbers 3, 5, 7, 9, 11, etc shows up once again.

Mathematicians have been fascinated by the connections between square and triangle numbers, and between ordinary numbers and numbers of special shape for centuries. They are still trying to find new connections between them.

Numbers ending in 5 can be squared mentally very easily: ignoring the 5 multiply the figures left by the number which follows it numerically, insert 25 after the product of this multiplication and you have the square of the original number. For example: $65^2 = 6 \times 7 = 42$, with 25 at the end; 4225; and $145^2 = 14 \times 15$, which is 210, with 25 on the end; 21025.

Numbers made up only of threes have a special pattern of squares:

$$\begin{aligned} 33^2 &= 1089 \\ 333^2 &= 110889 \\ 3333^2 &= 11108889 \\ 33333^2 &= 1111088889 \\ 333333^2 &= 111110888889 \end{aligned}$$

and so on.

Squares for other numbers can be found quickly by using the methods described in the chapter on multiplication.

## *FINDING SQUARE ROOTS*

This is the conventional method of finding square roots:

First divide the number up into groups by putting a mark over every second figure – starting with the units digit, going on to the hundreds digit and so on. For instance if the problem is to find the square root of 233289 set it out like this:

$$2\dot{3}3\dot{2}8\dot{9}$$

Now find the largest number the square of which is less than 23:

```
   2̇33̇28̇9(483
   16
   —
88)732
   704
   ———
963)2889
    2889
    ————
```

this is 4 – which gives you the first figure of the answer. Insert this to the right of the number, and subtract $4^2$ (16) from 23. Now bring down the next group of figures – 32 – and put it after the remainder just obtained to make 732.

The first part of the divisor of 732 (obtained by doubling 4, that part of the root you have already obtained) is 8. The final digit of the divisor, and the second of the root itself, is arrived at by trial. It will be the largest number that, inserted after the 8, and then used to multiply the number thus obtained, will give a product of less than 732. It turns out in this case to be 8 – for $8 \times 88 = 704$, and $732 - 704 = 28$.

Now go through the same process again:

Bring down the next group of numbers, 89, to make 2889.

Arrive at a divisor by doubling 48 (the part of the root already arrived at). This gives you 96.

Find by trial the next number of the square root (in this case the final one), which will also be the last digit of the divisor. It turns out to be 3, and you now have the complete root, 483, which can be checked by multiplying it by itself.

**Square roots by factors.** Square roots can also be found by breaking numbers down into factors which are the squares of known numbers (4, 9, 16, 25, and so on). Multiplying together the square roots of the factors will then give the square root of the number.

For instance the number 20736 is found to have the factors $4 \times 4 \times 4 \times 4 \times 9 \times 9$. The square root is then $2 \times 2 \times 2 \times 2 \times 3 \times 3$, which equals 144.

Both these methods are time consuming however – and there are shortcuts which, given a little practice, allow you to arrive at square roots mentally. First you must learn the table of squares of numbers from 1 to 9 by heart:

| *number* | *square* |
|---|---|
| 1 | 1 |
| 2 | 4 |
| 3 | 9 |
| 4 | 16 |
| 5 | 25 |
| 6 | 36 |
| 7 | 49 |
| 8 | 64 |
| 9 | 81 |

You will see from this table that any square ending in 1 will have a root ending in 1 or 9. A square ending in 4 will have a root ending in 2 or 8. A square ending in 9 will have a root ending in 3 or 7. A square ending in 6 a root ending in 4 or 6, and a square ending in 5 a root ending in 5. So any number that is a perfect square will end in 1, 4, 9, 6, or 5, and, with the exception of 5, the final number of the square will indicate two possible values for the last digit of the square root.

This table is the basis of a quick method of finding square roots. To find the root of a four-figure number first break the number into two groups of two digits. For example 6241 breaks into 62 and 41.

Now consider the group with the thousands and the hundreds digits. We know that the square root we are looking for will have two digits – to find the first we think where 62 stands in relation to the memorised table of squares. It is less than 64 but more than 49, so the highest possible first digit of the root is 7. We must now find the units digit of the square root. We know from our table that it must be 9 or 1 – for only they give squares ending in 1. To decide which take the number already arrived at as the first number of the root – in this case 7, multiply it by itself plus 1. If the product is *more* than the first two figures of the number for which you are finding the root take the lower of the two possible digits figures – in this case 1, to arrive at the square root, if it is *less* take the higher of the two possible figures – $7 \times (7 + 1) = 7 \times 8 = 56$, which is less than 62. The square root of 6241 is therefore 79. Here is another example.

To find the square root of 4096.

40 lies between 36 and 49, so the first figure of the square root is 6. The second figure could be 4 or 6. $6 \times 7 = 42$, which is more than 40 so we take the lower of the possible units digits. The square root is therefore 64.

When the square has five or six figures we know the square

root will have three. To find the square roots of numbers up to 40000 you must remember more squares – those up to 20:

| *number* | *square* |
|---|---|
| 11 | 121 |
| 12 | 144 |
| 13 | 169 |
| 14 | 196 |
| 15 | 225 |
| 16 | 256 |
| 17 | 289 |
| 18 | 324 |
| 19 | 361 |
| 20 | 400 |

With this table memorised you can set about finding the square root of longer numbers which are perfect squares. Let us take an example.

To find the square root of 15129.

First divide the number into two groups of digits by taking out the last two figures. This gives:

151   29

From the table you know that 151 falls between the squares of 12 and 13 – which are 144 and 169. The first two figures of the root you are extracting will therefore be 12 – the square root of the lower figure. To find the final figure you use the method already described: the number from which you are extracting the square root ends in a 9, the digit you are looking for must therefore be a 3 or a 7. To find which add 1 to 12 (the first figures of the root which you have already arrived at) and multiply the sum by 12. The product, 156, is greater than 151 so you take the lower of the two possible digits. You now have the complete square root of 15129, it is 123.

Here is another example.

To find the square root of 30276.

Divide the number into two groups of digits as before:

302 76

From the table we know that 302 falls between the squares of 17 and 18. The first two digits of the square root are therefore 17. The number ends in 6 so the final figure must be 4 or 6. $17 \times 18 = 306$, this is greater than 302, so you take the lower of the two possible digits to arrive at the square root of 174. If the number from which the square root is to be extracted is higher than 40000 you use a slightly different method.

For example, to extract the square root of 537289. First break it up into groups of two figures, starting from the right:

53 72 89

To find the hundreds figure of the square root go to the first memorised table. 53 stands between the squares of 7 and 8. The lower of these numbers is our first figure, 7.

To find the tens figure we must find the difference between 53 and the square of 7, that is between 53 and 49.

$$53 - 49 = 4$$

We now put 4 before the left-hand digit of the second group of figures in the number from which we are extracting the square. This gives 47. We divide this number by *twice* the figure already arrived at as the first in the square root – 7 – plus 1. This gives us 15

$$\frac{47}{15} = 3, \text{ with remainder } 2$$

(note that you are obtaining the quotient which will give the least remainder *above or below* – the remainder is the number that must be added to or subtracted from the dividend to bring it to the nearest multiple of the divisor).

The quotient thus obtained is the tens digit of the square root we are extracting, so the first two digits of the root are 73.

We know that the units digit must be 7 or 3, for the number ends in a 9. As the quotient obtained, 3, was greater than the remainder, 2, we take the smaller of the two possible figures and arrive at the complete square root of 733.

When the quotient is smaller than the remainder take the larger of the two possible figures as the units digit.

# 9

# *CUBES and CUBE ROOTS*

You will remember that square numbers could be illustrated by rectangular arrays of marbles. Cubes can be illustrated with square building blocks. You build a cube by adding layers of blocks to a square until there are as many layers of blocks as there are blocks in one side of the square.

A square made up of only 1 block is also 1 block deep. A square with a side of 2 blocks long will be made into a cube by adding another layer – there are 4 blocks in each layer and 8 in the whole cube.

$$2 \times 2 \times 2 = 8$$

A square with sides 3 blocks long contains 9 blocks, and a cube with sides of 3 blocks three times as many:

$$3 \times 3 \times 3 = 27$$

Going on up, enlarging the cube by one unit each time, we find the fourth cube contains 64 blocks, the fifth 125, and so on.

A cubic number is one obtained by multiplying a whole number by itself, and then by itself again. The cube root is the original number – the whole number you started with. Cubes are, as I said earlier, written with a superior 3. So 64 can be written as $4^3$. 4 on the other hand can also be written as $\sqrt[3]{64}$, that is to say the cube root of 64.

Another way of indicating the cube root of a number is by what is called a fractional index:

$$64^{\frac{1}{3}}$$

or, in words, 64 raised to the power of 1 over 3.

The cubic numbers, like the square numbers, have a special relation to the odd numbers:

$$\begin{aligned}
1^3 &= 1 &&= 1\\
2^3 &= 8 &&= 3 + 5\\
3^3 &= 27 &&= 7 + 9 + 11\\
4^3 &= 64 &&= 13 + 15 + 17 + 19\\
5^3 &= 125 &&= 21 + 23 + 25 + 27 + 29\\
6^3 &= 216 &&= 31 + 33 + 35 + 37 + 39 + 41\\
7^3 &= 343 &&= 43 + 45 + 47 + 49 + 51 + 53 + 55
\end{aligned}$$

and so on.

And cubic numbers, like square numbers, have a special relationship to triangular numbers: it is an odd one. If the cubic numbers are added up in order ($1^3$, $1^3 + 2^3$, $1^3 + 2^3 + 3^3$, and so on) the sums arrived at are the squares of the sequence of triangular numbers.

$$1^3 = 1 = 1^2$$
$$1 = \textit{triangle number } 1$$
$$1^3 + 2^3 = 9 = 3^2$$
$$3 = \textit{triangle number } 2$$
$$1^3 + 2^3 + 3^3 = 36 = 6^2$$
$$6 = \textit{triangle number } 3$$
$$1^3 + 2^3 + 3^3 + 4^3 = 100 = 10^2$$
$$10 = \textit{triangle number } 4$$
$$1^3 + 2^3 + 3^3 + 4^3 + 5^3 = 225 = 15^2$$
$$15 = \textit{triangle number } 5$$
$$1^3 + 2^3 + 3^3 + 4^3 + 5^3 + 6^3 = 441 = 21^2$$
$$21 = \textit{triangle number } 6$$
$$1^3 + 2^3 + 3^3 + 4^3 + 5^3 + 6^3 + 7^3 = 784 = 28^2$$
$$28 = \textit{triangle number } 7$$
$$1^3 + 2^3 + 3^3 + 4^3 + 5^3 + 6^3 + 7^3 + 8^3 = 1296 = 36^2$$
$$36 = \textit{triangle number } 8$$

and so on.

## *EXTRACTING CUBE ROOTS*

The most common conventional way of extracting cube roots is by factors – here, for example, is how you extract the cube root of 474552 by this method:

```
 2)474552
 --------
 2)237276
 --------
 2)118638
 --------
  3)59319
  -------
  3)19773
  -------
   3)6591
   ------
  13)2197
  -------
   13)169
   ------
    13)13
    -----
        1
```

The figure 474552 may thus be expressed as:

$$2^3 \times 3^3 \times 13^3$$

$$\sqrt[3]{474552} = \sqrt[3]{2^3 \times 3^3 \times 13^3} = 2 \times 3 \times 13 = 78$$

But there are simpler methods for extracting the roots of numbers which are exact cubes – I will show first how it can be done for numbers of up to six digits.

Again there is a table to remember, this time of the cubes of the numbers from 1 to 10:

| *number* | *cubes* |
|---|---|
| 1 | 1 |
| 2 | 8 |
| 3 | 27 |
| 4 | 64 |
| 5 | 125 |
| 6 | 216 |
| 7 | 343 |
| 8 | 512 |
| 9 | 729 |
| 10 | 1000 |

The first thing to notice about this table is that the numbers from 0 to 9 all occur. If a number is cubic one can therefore determine the units figure of the cube root immediately. If a cubic number ends in 1, 4, 5, 6, 9, or 0 its cube root will end in the same number. If it ends in 2 the cube root will end in 8, and if it ends in 8 the cube root will end in 2. Similarly if the number ends in 7 the units digit of the cube root will be 3, and if it ends in 3 the units digit of the cube root will be 7.

You will also need this table when you come to determine the tens digit.

If a cube has four, five, or six digits the root will contain two. To extract the root first divide the number into two groups by counting back three places from the units digit. For instance if you are extracting the cube root of 50653 divide it thus:

50 653

Find the first digit of the cube root by referring back mentally to the table of cubes. 50 falls between 27 and 64 – that is between $3^3$ and $4^3$: the tens figure of the cube root you are finding will be the root of the lower number – in this case 3.

Now you must find the units digit, and you know from the table that as the cubic number ends in 3 its cube root will end in 7.

The cube root you are seeking is therefore 37. Here is another example.

To find the cube root of 681472.

First divide the number into groups:

681 472

681 falls between the cubes of 8 and 9. The tens unit is therefore 8. The whole number ends in 2, the cube root therefore ends in 8. The cube root is 88.

To extract the cube root of longer cubic numbers – those with seven, eight or nine digits – is almost as simple once you have learnt yet another table:

| *X* | *Y* |
|---|---|
| 1 | 1 |
| 2 | 7 |
| 3 | 9 |
| 4 | 5 |
| 5 | 3 |
| 6 | 8 |
| 7 | 6 |
| 8 | 2 |
| 9 | 4 |
| 10 | 10 |

A number with seven, eight, or nine digits will have a cube root with three digits. To find the root we again begin by dividing the cubic number into groups. Moving from right to left make a break every three digits. For example to find the cube root of 92345408 first divide it thus:

92 345 408

To find the hundreds unit of the cube root think back to the table of cubes from 1 to 10. 92 falls between the cubes of 4 and 5, so 4 is the hundreds unit of the root we are extracting. The units digit – the terminal figure of the cube root – must, we know, be 2, for the last digit of the cubic number is 8.

Now we must find the tens digit. To do this first add up the odd digits of the original number – the first, third, fifth, seventh, and (if it is there) ninth figures in the number, reading from right to left. Subtract from this number the sum of the even figures in the number (the second, fourth, sixth, and eighth digits) added up in the same way. If the total of the even figures is greater than that of the odd figures add 11 to the latter before subtracting.

In the example this gives us:

$$8 + 4 + 4 + 2 = 18$$

(the sum of the odd digits) from which we subtract

$$0 + 5 + 3 + 9 = 17$$

(the sum of the even digits) to get a remainder of 1.

Now refer back to the table with columns $X$ and $Y$. Find in the $X$ column the number just obtained as a remainder and note the number opposite it in the $Y$ column. Add together the two digits of the final root which you already have (the hundreds and the units digits) and subtract from the sum the figure you have just obtained from the $Y$ column of the table.

In the example the remainder was 1, this stands opposite 1 in the table, so we subtract that figure from the sum of 4 and 2

$$4 + 2 = 6 \qquad 6 - 1 = 5$$

which gives us the tens unit of the cube root. The complete root is 452.

If the sum of the hundreds and units digits is less than the figure obtained from column $Y$ of the table add 11 to that sum before making the subtraction. Here is another example.

Find the cube root of 665338617.

Divide the number into groups:

$$665 \quad 338 \quad 617$$

The hundreds digit of the root is 8 as 665 stands between the cubes of 8 and 9.

The units digit is 3 as the number ends in 7.

The sum of the digits in the odd places is

$$7 + 6 + 3 + 5 + 6 = 27$$

from which is subtracted the sum of the digits in the even places

$$27 - (1 + 8 + 3 + 6) = 27 - 18 = 9$$

In the table, 9 in the $X$ column stands opposite 4 in the $Y$ column. The sum of the digits already arrived at $(8 + 3 = 11)$ minus the digit from the $Y$ column, gives us the tens unit of the cube root $(11 - 4 = 7)$.

The cube root is therefore 873.

## *EXTRACTING FIFTH ROOTS*

When a number has been raised to the fifth power the root is surprisingly easy to find.

If the fifth power has between six and ten digits the root will have two.

The units digit of all fifth roots is the same as the units digit of the fifth power of the number, so all you have to find is the tens digit. To do this divide the number at the fifth place, reading from right to left. For example if the number you are finding the fifth root of is 33554432 divide it thus:

335 54432

For the next step you need another table – that of fifth powers of numbers from 1 to 10:

| *number* | *fifth power* |
|---|---|
| 1 | 1 |
| 2 | 32 |
| 3 | 243 |
| 4 | 1024 |
| 5 | 3125 |
| 6 | 7776 |
| 7 | 16807 |
| 8 | 32768 |
| 9 | 59049 |
| 10 | 100000 |

When you have committed this to memory you can quickly determine where the left-hand group fits in the table. In the example 335 comes between $3^5$ and $4^5$, the fifth root of 33554432 is therefore 32; here as with square and cube roots the root of the lower of the pair of powers that bracket the value of the first group in the table gives you the first figure of the root you are seeking. Here is another example.

Find the fifth root of 1073741824.

First divide at the fifth place:

10737 41824

10737 comes between the fifth powers of 6 and 7 in the table, the fifth root is therefore 6 followed by the final digit of the original number: 64.

By similar methods other roots can be found. Ninth, thirteenth, seventeenth, twenty-first and twenty-fifth roots are like fifth roots in that the power and the root have the same terminal digit – you can extract the root therefore by constructing and memorising tables of powers from 1 to 10. Other roots – fourth, sixth, and so on, can be extracted by the same methods as those described here for square, cube, and fifth roots; try working some of these out for yourself.

# 10

# *PERCENTAGES, DISCOUNTS and INTEREST*

Percentages – fractions expressed as parts per 100 – are particularly useful in money sums: rates of interest and discounts are both calculated in this way. There are a number of shortcuts that can be used to simplify percentage sums and make them easier to do mentally.

For instance the problem may be to discount \$400 by 30 per cent. The normal calculation runs as follows:

$$\frac{400 \times 30}{100} = 120$$

Deduct 120 (30 per cent of 400) from 400 to arrive at the answer:

$$400 - 120 = \$280$$

A quicker method, that can be done mentally, is to deduct 30 per cent from 100 – which gives you 70. 70 per cent of \$100 is \$70, so 70 per cent of \$400 is four times as much – \$280.

Even if you are not working with round figures, and the amount to be discounted has cents as well as whole dollars (or pence as well as pounds, or paise as well as rupees – the basic calculations are of course the same for any decimal currency) the method of arriving directly at the discounted price, rather than finding the amount of the discount and deducting it, works.

For instance to find the discounted price of a sewing machine that is offered at a discount of 30 per cent on the marked price of \$185·60 you must calculate 70 per cent of that sum. This involves multiplying by 70 and dividing by 100 – or, (having divided both by 10) multiplying by 7 and dividing by 10.

To divide by 10 shift the decimal point one place to the left; you can then arrive at the answer by a simple multiplication by 7:

$$18{\cdot}56 \times 7 = 129{\cdot}92$$

which gives the answer of \$129·92 directly.

Similarly, to find \$496·20 less 20 per cent, multiply 49·62 by 8 to arrive at the answer of \$396·96.

To find what remains when a sum is discounted by 10 per cent move the decimal point one place to the left in the original amount and subtract the amount thus arrived at from the original sum: for example to discount \$28·50 by 10 per cent deduct \$2·85 to arrive at the discounted price of \$25·65.

A 90 per cent discount leaves 10 per cent of the original price – a simple shift of the decimal point one place to the left gives the answer.

When more complicated percentages are being worked out it is often simpler to break the figures down. For instance to find 32½ per cent add 25 per cent to 5 per cent to 2½ per cent. By this method you would arrive at 32½ per cent of \$280 in the following way:

| | |
|---|---|
| 25 per cent of 280 | = 70 |
| 10 per cent = 28 so | |
| 5 per cent of 280 | = 14 |
| $2\frac{1}{2}$ per cent of 280 (half of 5 per cent) | = 7 |
| $32\frac{1}{2}$ per cent of 280 | = 91 |

Similarly to calculate $67\frac{1}{2}$ per cent of $460:

| | |
|---|---|
| 50 per cent of 460 | = 230 |
| 10 per cent of 460 | = 46 |
| 5 per cent of 460 (half of 10 per cent) | = 23 |
| $2\frac{1}{2}$ per cent of 460 (half of 5 per cent) | = 11·50 |
| $67\frac{1}{2}$ per cent of 460 | = 310·50 |

The break down can be done in various ways – the sum above, for instance, could be done like this:

| | |
|---|---|
| 50 per cent of 460 | = 230 |
| $12\frac{1}{2}$ per cent of 460 (quarter of 50 per cent) | = 57·50 |
| 5 per cent of 460 (tenth of 50 per cent) | = 23 |
| $67\frac{1}{2}$ per cent of 460 | = 310·50 |

Here are two more examples – if the fractions are carefully chosen quite complicated percentages turn into simple additions of parts, or parts of parts.

| | |
|---|---|
| 25 per cent of 298 | = 74·50 |
| $12\frac{1}{2}$ per cent of 298 (half of 25 per cent) | = 37·25 |
| 5 per cent of 298 (fifth of 25 per cent) | = 14·90 |
| $42\frac{1}{2}$ per cent of 298 | = 126·65 |

To find 52½ per cent of $368:

| | | |
|---|---|---|
| 50 per cent of 368 | = | 184 |
| 2½ per cent of 368 (twentieth of 50 per cent) | = | 9·20 |
| 52½ per cent of 368 | = | 193·20 |

The general principle can be applied to any percentage sum – you will find ways of doing most of them in your head. But mistakes in money sums are likely to be costly so check the result by reversing the calculation – having found what 42½ per cent of a given sum is, find what sum your answer would be 42½ per cent of.

## *SIMPLE INTEREST*

Interest paid on a loan or a deposit is also expressed as a percentage. This states the amount due each year – the amount to be paid by the borrower to the lender. When less than a whole year is involved, the amount due is a fraction of the yearly sum. Ordinary interest is arrived at by assuming there are 360 days in the year and writing the fraction as the number of days for which interest is due over 360. Exact interest is determined on a 365-day year, and is used where large sums are involved and generally by banks and financial institutions.

The usual way of doing interest sums is this: multiply the number of days of the interest period over 360 or 365 by the amount of the loan by the interest rate over 100. For instance the ordinary interest on $326 for a period of 60 days at 6 per cent would be worked out as follows:

$$\frac{60 \times 326 \times 6}{360 \times 100}$$

Cancelling out simplifies the sum to

$$\frac{326}{100} = 3{\cdot}26$$

This is the interest due.

The 6 per cent and 60 days, and the 360 days of the ordinary interest year cancel out neatly. This fact can, in less obvious interest calculations, help you to arrive at methods of doing them mentally. For instance to find the ordinary interest on $398 for 70 days at 6 per cent (the calculations are to nearest whole penny):

| | |
|---|---|
| Interest for 60 days at 6 per cent | = 3·98 |
| Interest for 10 days (sixth of 60 days) | = ·66 |
| Interest for 70 days | = 4·64 |

Or to find the ordinary interest on $426 for 36 days at 6 per cent:

| | |
|---|---|
| Interest for 30 days (half of 60 days) | = 2·13 |
| Interest for 6 days (tenth of 60 days) | = ·42 |
| Interest for 36 days | = 2·55 |

Knowing that ordinary interest of 6 per cent for 60 days is always one-hundredth of the principal makes all interest sums of this kind straightforward. When the rate of interest is other than 6 per cent the method can sometimes be adapted.

For instance to find the interest due on $284 for 54 days at 4½ per cent:

Interest for 60 days at 6 per cent $= 2{\cdot}84$
Interest for 60 days at $4\frac{1}{2}$ per cent $= 2{\cdot}13$
(2·84 less a quarter)
Interest for 6 days at $4\frac{1}{2}$ per cent $= {\cdot}21$

Interest due for 54 days is therefore \$2·13 less 21 cents – that is \$1·92.

## *COMPOUND INTEREST*

If the interest due on a loan is not paid, but allowed to accumulate, the sum invested will be larger at the end of each interest period. When the interest and the original sum are 'compounded' in this way the amount due at the end of each interest period rises – for interest is not only paid on the original sum, but on the interest already earned as well. Interest calculated in this way is called compound interest. Compound interest, like simple interest, is calculated by multiplying the principal by the period by the interest rate. In the case of compound interest, however, the operation must be repeated for each interest period.

If \$3000 is lent for a year at 4 per cent simple interest, payable quarterly the interest due at the end of each period will be \$30, and the total interest earned over the year \$120. Over two years it will earn twice that – \$240. Now let us see what happens if the loan is for \$3000, but at 4 per cent compound interest, computed quarterly, and what is earned over the same period of two years.

| | |
|---|---|
| Principal | 3000·00 |
| Interest for 1st quarter at 1 per cent | 30·00 |
| Principal at the beginning of 2nd quarter | 3030·00 |
| Interest for the period of 2nd quarter | 30·30 |
| Principal at the beginning of 3rd quarter | 3060·30 |
| Interest for the period of 3rd quarter | 30·61 |
| Principal at the beginning of 4th quarter | 3090·91 |
| Interest for the period of 4th quarter | 30·90 |
| Principal at the beginning of 5th quarter | 3121·81 |
| Interest for the period of 5th quarter | 31·21 |
| Principal at the beginning of 6th quarter | 3153·02 |
| Interest for the period of 6th quarter | 31·53 |
| Principal at the beginning of 7th quarter | 3184·55 |
| Interest for the period of 7th quarter | 31·84 |
| Principal at the beginning of 8th quarter | 3216·39 |
| Interest at the end of 8th quarter | 32·16 |
| TOTAL principal at the end of two years | 3248·55 |

The total interest earned is \$3248·55 — \$3000, that is \$248·55. The difference between simple and compound interest in this case is therefore \$8·55.

There is no straightforward method of calculating compound interest mentally, alas, although the shortcuts I have already described for working with percentages will often make the calculation quicker.

# 11

# DECIMALS

Decimals, as I explained in Chapter One, are one way of expressing parts of whole numbers. To the left of the decimal point the value of each place is ten times greater than the value of the place to its right. To the right of the decimal point each place is one-tenth of the value of the place to its left.

| millions | hundred thousands | ten thousands | thousandths | hundreds | tens | units | | tenths | hundredths | thousands | ten thousandths | hundred thousandths | millionths | ten millionths |
|---|---|---|---|---|---|---|---|---|---|---|---|---|---|---|
| 7 | 6 | 5 | 4 | 3 | 2 | 1 | · | 1 | 2 | 3 | 4 | 5 | 6 | 7 |

16·3 would be read as 16 and 3 tenths
16·38 would be read as 16 and 38 hundredths
16·384 would be read as 16 and 384 thousandths
16·3846 would be read as 16 and 3846 ten thousandths
16·38468 would be read as 16 and 38468 hundred thousandths
16·384681 would be read as 16 and 384681 millionths

To shift the decimal point one place to the left is to divide by 10.

To shift it one place to the right is to multiply by 10. This is at once very simple and a potential source of errors – a confusion of one place about where the decimal point should go will give you an answer too small or too great by a factor of 10. The rules though are simple.

**In adding and subtracting** see that the decimal points stand directly below one another. The decimal point in the answer will come in the same place:

```
 39·3
  9·0633
  1·5161
  3·03
 46·8
 ———————
 99·7094
```

In subtracting you may have to add zeros to make up the number being subtracted from to the same number of places as the number being subtracted – or vice versa:

9·36 from 41·0853421 becomes

```
 41·0853421
  9·3600000
 ——————————
 31·7253421
```

or 9·364857 from 12·62 becomes

```
 12·620000
  9·364857
 —————————
  3·255143
```

**In multiplying** carry out the calculation first and then find the place to insert the decimal point by counting as many places in from the right in the product as there are places in the multiplier and multiplicand together.

$$\begin{array}{r} 9{\cdot}4562 \\ \times 7{\cdot}0956 \\ \hline 67{\cdot}09741272 \end{array}$$

There are four decimal places in the multiplier, and in the multiplicand, and eight therefore in the product. When multiplying very small decimal fractions the zeros after the decimal point can be ignored until the final process of inserting the point in the product:

$$\begin{array}{r} {\cdot}000056 \\ \times {\cdot}0923 \\ \hline \end{array}$$

First of all multiply 56 by 923 to obtain the product 51688. In the multiplicand there are six digits following the decimal point, and in the multiplier four digits. Therefore the product is ·0000051688.

The same principle applies when multiplying more than two numbers together: to multiply 3·15 by ·315 by 31·5 first find the product of 315 × 315 × 315, which is 31255875 and count off six decimal places from the right – the total number of places in the three numbers being multiplied together.

**In division** too the same technique is used as with whole numbers. The placing of the decimal point is, however, easier if the divisor or dividend is a whole number. This can be achieved by shifting the decimal point to the right. Provided the point is moved the same number of places in both numbers the result of the division will be the same. For example to divide ·7 by ·196:

First multiply both the dividend and the divisor by 10. As both the numbers are multiplied by the same number, the quotient remains unaltered. ·7 becomes 7 and ·196 becomes 1·96. Now carry out the division:

```
1·96)7·00(3·57
     5·88
     ----
     1·120
       980
     -----
      1400
      1372
      ----
        28
```

And so on.

More generally the best rule is to move the decimal point to the right in the divisor as many places as are necessary to make it a whole number. Move the decimal point the same number of places to the right in the dividend. Insert the decimal point in the quotient as soon as any figure is used from the decimal part of the dividend in working out the problem.

Some fractions cannot be expressed as decimals with a finite number of places; among these are fractions like 1/3 (·3333333 etc) and 1/7 (·142857142857142857 etc) which have single figures or groups of figures recurring infinitely. But it is not only in cases like this that some decimal places are ignored. When starting a calculation you should decide what degree of accuracy is required. It is not necessary to continue any operation beyond the third or fourth decimal place in most practical calculations. To determine that a number is correct to a certain number of decimal places you must look at the rejected figures. If the result is to be correct to four places you must know what number stands in the fifth place. If it is 5 or greater add one to the preceding digit. For example in the multiplication

$$
\begin{array}{r}
13{\cdot}064973 \\
\times\ 25{\cdot}035\phantom{000} \\
\hline
327{\cdot}081599055
\end{array}
$$

the answer runs to nine places of decimals. This, correct to one place of decimals is 327·1, for 8 is greater than 5 and 1 is added to the preceding figure. Correct to two places the answer 327·08, and to three places 327·082. Correct to four places it is 327·0816.

As the easiest mistake to make with calculations involving decimals is putting the decimal point in the wrong place it is always as well to make sure that you have a rough idea of the magnitude of the expected answer. In an earlier example we multiplied 3·15 by ·315 by 31·5. This is, very roughly, 3 × 1/3 × 30, 1/3 of 3 is 1 and the answer should have two significant figures in the region of 30, followed by decimal places. Having made this rough check you can have confidence in the answer 31·255875 when you arrive at it.

# 12

# *FRACTIONS*

In the definitions at the beginning of this book I described fractions. You will remember that a fraction has a numerator – which tells how many parts you have – and a denominator – which tells what proportion of a whole those parts are. Thus in the fraction 3/4 the 3 tells how many parts there are, and the 4 that each is a quarter of a whole. If the numerator and denominator are both multiplied by the same figure the resulting fraction will have the same value as the original one – 3/4 or 6/8 or 9/12 are all the same proportion of a whole. It is usually easiest to deal with fractions reduced to their lowest terms, however, and if numerator and denominator have a common factor both are divided by it to achieve this reduction. 9/27 for instance is divided by the largest common factor of 9 and 27, which is 9, to arrive at the lowest terms of the fraction, 1/3.

*Common fractions* – fractions with a numerator and a denominator are either *proper fractions* – those which have a larger denominator than numerator, *improper fractions* – those in which the numerator is greater than or the same as the denominator or *mixed numbers* – which consist of a proper fraction and a whole number. Any improper fraction can also be written as a mixed number and any mixed number as an improper fraction, $2\frac{3}{4}$ for instance as 11/4.

**Adding fractions.** To add fractions we must find a common denominator. That is to say we must add quarters to quarters and eighths to eighths, not quarters to eighths. The simplest way to find a common denominator is to multiply the denominators together. We know that a fraction is not changed when its numerator and denominator are divided or multiplied by the same figure, so both numerator and denominator can, for each item in the addition be multiplied by the product of the other denominators. For example:

$$\frac{1}{2}+\frac{1}{3}+\frac{2}{5}+\frac{1}{4}$$

The common denominator in this problem is the product of all the denominators in the original fractions. That is:

$$2 \times 3 \times 5 \times 4 = 120$$

Thus the problem is represented as:

$$\frac{60}{120}+\frac{40}{120}+\frac{48}{120}+\frac{30}{120}=\frac{60+40+48+30}{120}$$

$$=\frac{178}{120}=\frac{89}{60}=1\frac{29}{60}$$

You will remember from Chapter 7 the procedures for finding the least common multiple of a group of numbers. The most convenient common denominator will be the least common multiple of the denominators of the fractions being added or subtracted. In the example above for instance we could have used 60 rather than 120 as the common denominator. If we add

$$\frac{3}{4}+\frac{1}{3}+\frac{1}{6}+\frac{7}{12}$$

the common denominator is 12, as it is the biggest denominator and each of the other denominators are exactly divisible into it. We change the other denominators to 12 and the problem becomes:

$$\frac{9}{12}+\frac{4}{12}+\frac{2}{12}+\frac{7}{12}=\frac{9+4+2+7}{12}$$
$$=\frac{22}{12}=\frac{11}{6}=1\frac{5}{6}$$

When subtracting fractions you go through the same process of finding a common denominator and multiplying the numerators. For instance to subtract 9/11 from 1/2: $2 \times 11 = 22$ which is your common denominator. The problem then becomes

$$\frac{18}{22}-\frac{11}{22}=\frac{7}{22}$$

And addition and subtraction can of course be combined:

$$\frac{1}{2}+\frac{3}{5}+\frac{1}{3}-\frac{3}{4}$$

Here the common denominator is $2 \times 5 \times 3 \times 4 = 120$. Therefore the problem becomes:

$$\frac{60}{120}+\frac{72}{120}+\frac{40}{120}-\frac{90}{120}=\frac{60+72+40-90}{120}$$
$$=\frac{82}{120}=\frac{41}{60}$$

When adding and subtracting mixed numbers it is less confusing if you deal with the whole numbers first, and then the fractions. For example $6\frac{1}{4} + 2\frac{1}{2} + 1\frac{1}{8} + 3\frac{1}{8}$ becomes:

$$6+2+1+3+\frac{1}{4}+\frac{1}{2}+\frac{1}{8}+\frac{1}{8}=12+\frac{1}{4}+\frac{1}{2}+\frac{1}{8}+\frac{1}{8}$$
$$=12+\frac{2}{8}+\frac{4}{8}+\frac{1}{8}+\frac{1}{8}$$
$$=12+\frac{2+4+1+1}{8}$$
$$=12+\frac{8}{8}$$
$$=12+1=13$$

When **multiplying** by fractions the numerator of the product is obtained by multiplying the numerators of the two fractions and the denominator of the product by multiplying the denominators. The product will of course be smaller than the two original fractions – $\frac{1}{4}$ of $\frac{1}{2}$ is $\frac{1}{8}$. The method is very simple. For example to find the product of $5/8 \times 6/9$:

$$\frac{5\times 6}{8\times 9}=\frac{30}{72}=\frac{5}{12}$$

To **divide** one fraction by another – to find for instance how many eighths there are in a half – invert the fraction you are dividing by – the divisor – and multiply the numerators and denominators together, as in multiplication. For example 3/7 divided by 6/9:

$$\frac{3}{7}\times\frac{9}{6}=\frac{3\times 9}{7\times 6}=\frac{27}{42}=\frac{9}{14}$$

In multiplications and divisions the work can be simplified if, whenever possible you cancel – that is divide a numerator and a denominator by the same number. For example 3/4 × 8/9 becomes 1/1 × 2/3, or 2/3 when the 4 and the 8 and the 3 and the 9 have been cancelled.

Cancelling can be done in more than one step: a numerator may have different factors in common with different denominators and vice versa. It can of course only be done when all the signs in the calculation are multiplication signs.

# 13

# *THE CALENDAR*

In this chapter I will show you how to determine what day of the week any date fell on or will fall on. Any date, that is, since 15 October 1582 when our present calendar – the Gregorian calendar – was instituted.

In order to do this mentally you will need to commit four tables to memory. The first is so simple it hardly justifies the name. Here it is:

**TABLE I**

| 7 | 14 | 21 | 28 |
|---|---|---|---|

The second is more complicated:

**TABLE II**

| | |
|---|---|
| *January* | *corresponds to* 0 |
| *February* | *corresponds to* 3 |
| *March* | *corresponds to* 3 |
| *April* | *corresponds to* 6 |
| *May* | *corresponds to* 1 |
| *June* | *corresponds to* 4 |

*July* *corresponds to* 6
*August* *corresponds to* 2
*September* *corresponds to* 5
*October* *corresponds to* 0
*November* *corresponds to* 3
*December* *corresponds to* 5

The third is longer still:

TABLE III

1900 *corresponds to* 0
1904 *corresponds to* 5
1908 *corresponds to* 3
1912 *corresponds to* 1
1916 *corresponds to* 6
1920 *corresponds to* 4
1924 *corresponds to* 2
1928 *corresponds to* 0
1932 *corresponds to* 5
1936 *corresponds to* 3
1940 *corresponds to* 1
1944 *corresponds to* 6
1948 *corresponds to* 4
1952 *corresponds to* 2
1956 *corresponds to* 0
1960 *corresponds to* 5
1964 *corresponds to* 3
1968 *corresponds to* 1
1972 *corresponds to* 6
And so on.

Your starting point will, of course be a date – a specific day of a given month of a given year.

At the end of your calculation you will have a figure which will correspond to a day of the week in the following way:

TABLE IV

| | |
|---|---|
| *Sunday* | 0 |
| *Monday* | 1 |
| *Tuesday* | 2 |
| *Wednesday* | 3 |
| *Thursday* | 4 |
| *Friday* | 5 |
| *Saturday* | 6 |

To arrive at this final figure you carry out the following operations:

1. Reduce the date of the month to a figure less than 7 by subtracting the appropriate multiple of 7 from Table I.

2. To the figure thus obtained add the figure standing opposite the relevant month in Table II. Again cast out the sevens.

3. If the year falls after 1900 and is a leap year add the figure obtained after operation 2 to the figure standing opposite the year in Table III, *unless* the month is January or February, in which case add one less than the figure shown. If the year is not a leap year add to the figure obtained after operation 2 the figure standing opposite the previous leap year, plus the difference between that leap year and the year in question. (If, for instance, the year in question is 1942 add to 1 – which is the number standing opposite 1940 – the difference between 1940 and 1942, to arrive at the figure to add to the total obtained after operation 2.)

4. Having arrived at the total of date, month, and year figures cast out the sevens again (as in operation 1) to arrive at the figure which will correspond to the required day of the week in Table IV.

The operations described here will give you the day of the week for any date in the 20th century. (Remember by the way that 1900 is not a leap year, so there is no need to reduce by one unit for January or February.) To find what day dates between 1582 and the beginning of the 20th century fell on you need to add a century figure to the figure obtained for the equivalent year in the 20th century. For the 19th century this is 2, for the 18th century it is 4 and for the 17th century 6. Here are some examples.

**What day of the week was 9 April 1932?**

First 9 is reduced by subtracting 7. The date figure is therefore 2. In Table II the figure corresponding to April is 6. This is added to 2.

$$2 + 6 = 8$$

Cast out the 7, and you are left with 1. From Table III you find that 1932 stands against 5.

$$5 + 1 = 6$$

From Table IV you get the day of the week you are looking for. The number 6 signifies Saturday; 9 April 1932 was a Saturday.

**What day of the week was 23 January 1940?**

Reduce 23 by casting out the sevens:

$$23 - 21 = 2$$

Add the number for the month of January:

$$2 + 0 = 2$$

Add the number for the year:

$$2 + 1 = 3$$

but in this case it is leap year and the month is January so subtract 1:

$$3 - 1 = 2$$

which, by reference to Table IV, we find stands for Tuesday: 23 January 1940 was a Tuesday.

**What day of the week was 19 October 1942?**

Reduce 19 by casting out the sevens:

$$19 - 14 = 5$$

Add the number for the month of October:

$$5 + 0 = 5$$

The leap year before 1942 is 1940. This has the number 1, which you add to the difference between 1940 and 1942:

$$1 + 2 = 3$$

This is added to the result of the previous operation:

$$3 + 5 = 8$$

We cast out the sevens, to get the answer 1, which tells us that 19 October 1942 was a Monday.

Table III can be extended into the future to give days of the week for any year in this century. We will now look at some examples going further back – to work these out we will need the century figures, as well as date, month and year numbers.

**What day of the week was 9 August 1832?**

| | |
|---|---|
| The date figure (9 – 7) | 2 |
| The month figure | 2 |
| The figure corresponding to the same year in the present century | 5 |
| The century figure | 2 |
| Total | 11 |

Casting out the sevens gives 4; 9 August 1832 was a Thursday.

**What day of the week was 2 October 1869?**

| | |
|---|---|
| Date figure | 2 |
| Month figure | 0 |

| | |
|---|---|
| Year figure (the figure for the nearest leap year, '68, plus the difference between '68 and '69) | 2 |
| Century figure | 2 |
| Total | 6 |

2 October 1869 was a Saturday.

**What day of the week was 29 July 1744?**

| | |
|---|---|
| Date figure (29 – 28) | 1 |
| Month figure for July | 6 |
| Figure of '44 | 6 |
| Figure for 18th century | 4 |
| Total | 17 |

Casting out the sevens leaves 3; 29 July 1844 was a Wednesday.

**What day of the week was 26 December 1613?**

| | |
|---|---|
| Date figure (26 – 21) | 5 |
| Month figure for December | 5 |
| Year figure for '13 (1 + 1) | 2 |
| Figure for 17th century | 6 |
| Total | 18 |

Casting out the sevens leaves 4; 26 December 1613 was a Thursday.

Once you have the tables by heart the calculations can be done mentally at very great speed.

## *EXACT AGES*

If you know a person's date of birth it should be very easy to find out how old he is in years, months and days. And sometimes it is, for the subtraction can be done like this:

**What is the exact age of a person on 12 May 1973 if he was born on 8 February 1936?**

|  |  |  |
|---|---|---|
| 1973 | 5 | 12 |
| 1936 | 2 | 8 |
| 37 | 3 | 4 |

However, when there are more days in the date being subtracted than in that being subtracted from, days have to be borrowed. A month is taken from the bottom line and added to the top line. For example:

**What is the exact age of a person on 12 May 1973 if he was born on 17 March 1942?**

|  |  |  |
|---|---|---|
| 1973 | 5 | 12 |
| 1942 | 3 | 17 |

17 is greater than 12 so days must be added to the top line. The third month, March, has 31 days, so 17 is subtracted from $31 + 12$, to give the number of days – 26. Before subtracting the months 1 must be added to the bottom line.

The answer is 31 years 1 month and 26 days. Or again:

**What is the exact age of a person on 24 December 1972 if he was born on 30 August 1939?**

|  |  |  |
|---|---|---|
| 1972 | 12 | 24 |
| 1939 | 8 | 30 |
| 33 | 3 | 25 |

(31 is added to 24 in the top line in the days column, and 1 therefore to the months in the bottom line.)

When the number of months in the top line is smaller than the number of months in the bottom line add 12 to the top line, and 1 to the years in the bottom line. For example:

**What is the exact age of a person on 19 August 1973 if he was born on 16 October 1942?**

| | | |
|---|---|---|
| 1973 | 8 | 19 |
| 1942 | 10 | 16 |
| 30 | 10 | 3 |

(12 is added to the months column in the top line, and 1 to the years in the bottom line.) Or again:

**What is the exact age of a person on 23 September 1973 if he was born on 29 November 1947?**

| | | |
|---|---|---|
| 1973 | 9 | 23 |
| 1947 | 11 | 29 |
| 25 | 9 | 24 |

(30 is added to the days in the top line, and 1 to the months in the bottom line, 12 is added to the months in the top line and 1 to the years in the bottom line.)

# 14

# *SOME SPECIAL NUMBERS*

In Chapter Two, when I was writing about the digits, I described some of their special characteristics – the palindromes created by the multiplication of 1, 11, 111 etc by themselves for instance, or the pattern thrown up when one adds together the digits of products of a series of numbers multiplied by 4.

Here I will say something more about such characteristics. The personalities of numbers – and of families of numbers – are such that there are still mysteries to be probed.

In writing of the number 6 in Chapter Two I said it was one of the group of **perfect numbers.** Perfect numbers are those that are equal to the sum of all their divisors that are less than themselves. This is of course not common: $1 + 2 + 3 = 6$, but $1 + 3$ is less than 9, and $1 + 2 + 3 + 4 + 6$ is greater than 12. In fact the second perfect number is 28:

$$1 + 2 + 4 + 7 + 14 = 28$$

and for the third you must go on to 496:

$$1 + 2 + 4 + 8 + 16 + 31 + 62 + 124 + 248 = 496$$

The fourth perfect number is 8128:

$$1 + 2 + 4 + 8 + 16 + 32 + 64 + 127 + 254 + 508 + 1016 + 2032 + 4064 = 8128$$

There was a gap of 1400 years between the discovery of the fourth and fifth perfect numbers. The fifth is 33550336 and the sixth 8589869056. Even today there are only 17 perfect numbers known – the seventeenth has 1373 digits.

Another family of numbers I have already touched on are the **triangle numbers**. You will remember that it was a characteristic of square numbers that they were equal to the sum of two adjoining members of the sequence of triangle numbers. Think of triangle numbers as being built up from rows of marbles. The first triangle number is 1 – one row, one marble. As each new row has one more marble than the row before it the next triangle number is 3 – one marble in the first row, two in the second. The number of marbles in each new row is the same as the number of rows in the triangle – and any triangle number is the sum of a sequence of digits from one upwards. The first triangle number is labelled T1, the second T2 and so on. Here are the first seven of them:

| | | |
|---|---|---|
| T1 | (1) | = 1 |
| T2 | (1 + 2) | = 3 |
| T3 | (1 + 2 + 3) | = 6 |
| T4 | (1 + 2 + 3 + 4) | = 10 |
| T5 | (1 + 2 + 3 + 4 + 5) | = 15 |
| T6 | (1 + 2 + 3 + 4 + 5 + 6) | = 21 |
| T7 | (1 + 2 + 3 + 4 + 5 + 6 + 7) | = 28 |

It will be seen that to find T10 or T22 we must add the sequence of digits from 1 to 10 or from 1 to 22. This is obviously laborious and there is a quick method of finding the value of any given T number. To understand how it works think back to the pattern made by the marbles as we built up the triangle numbers. Here for instance is T4:

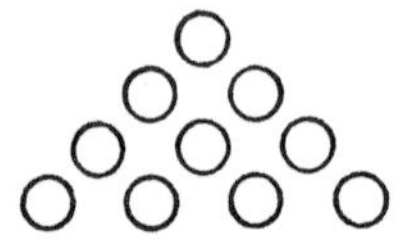

Turn this triangle through 45 degrees and add another to it to form a rectangle:

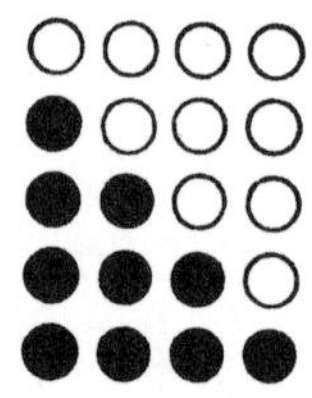

One side of the rectangle has the same number of marbles as the sides of the triangle. The other has that number plus 1. The number of marbles in the rectangle can be found by multiplying the number of marbles on the short side by the number on the long side – in the example 4 × 5. But the rectangle was made up from two identical triangles, and the number of marbles in each can be obtained by dividing the total number of marbles in the rectangle by 2. This principle can be seen to work for all triangle numbers. The T number gives the number of rows: the rectangle will have one side with the same number of marbles as the T number, the adjacent side will have one more. The product of the T number and the T number plus 1, when divided by 2, gives the value of the T number in question. For instance to find the value of T15:

$$\frac{15 \times 16}{2} = 120$$

And to find T20:

$$\frac{20 \times 21}{2} = 210$$

Or T236:

$$\frac{236 \times 237}{2} = 27966$$

## *142857 – THE REVOLVING NUMBER*

I described some of the oddities of 142857 when discussing the digit 7 in Chapter Two. Its habit of repeating itself in the total when the powers of 2 are multiplied by 7 and added, after being set out in a staggered formation, for instance. Here are some more.

First set out the digits in a circle:

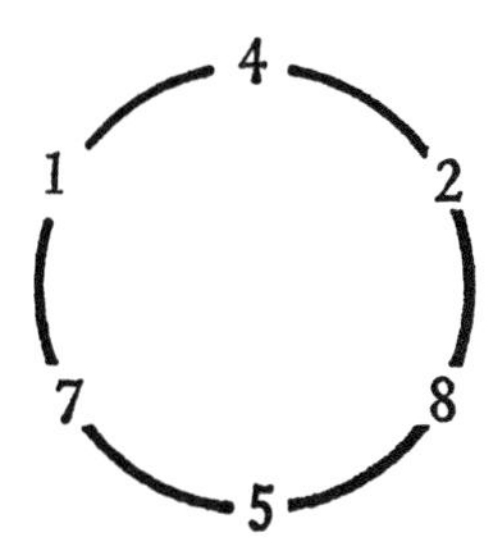


Now multiply 142857 by the numbers from 1 to 6:

| 142857 | 142857 | 142857 | 142857 | 142857 | 142857 |
|---|---|---|---|---|---|
| ×1 | ×2 | ×3 | ×4 | ×5 | ×6 |
| 142857 | 285714 | 428571 | 571428 | 714285 | 857142 |

You will see that the number starts revolving – the same digits in different combinations arrived at by starting from a different point on the circle. Multiply 142857 by 7 and things suddenly change:

$$\begin{array}{r} 142857 \\ \times 7 \\ \hline 999999 \end{array}$$

But there are still oddities in store, for instance if you multiply 142857 by a really big number see what happens.

$$\begin{array}{r} 142857 \\ \times\ 32284662474 \\ \hline 4612090027048218 \end{array}$$

No resemblance to 142857 at first sight perhaps, but divide the product up into groups of 6, 6, and 4 and see what happens:

$$\begin{array}{r} 048218 \\ 090027 \\ 4612 \\ \hline 142857 \end{array}$$

The revolving number appears again! Sometimes 142857 hides deeper than that. For instance in this multiplication the product seems immune from 142857 extraction:

$$\begin{array}{r} 142857 \\ \times\ 45013648 \\ \hline 6430514712336 \end{array}$$

But divide it into 6, 6, and 1 in the same way as you did last time and you arrive at the following sum:

$$\begin{array}{r} 712336 \\ 430514 \\ 6 \\ \hline 1142856 \end{array}$$

Treat this total in the same way:

$$\begin{array}{r} 142856 \\ 1 \\ \hline 142857 \end{array}$$

And you winkle 142857 out.

You can find oddities just by looking at the digits themselves. For instance if you divide them into two groups, 142 and 857, the second figure of the first group, multiplied by the third figure of the first group gives the first figure of the second group:

$$4 \times 2 = 8$$

The sum of the first two figures of the first group gives the second figure of the second group:

$$1 + 4 = 5$$

And the sum of all the three figures of the first group gives the third figure of the second group:

$$1 + 4 + 2 = 7$$

To show the next property of 142857 you have to draw up a table of the products of the number when multiplied by 1, 2, 3, 4, 5, and 6. Horizontally and vertically the digits all add up to 27:

| | | | | | | | |
|---|---|---|---|---|---|---|---|
| 1 | 4 | 2 | 8 | 5 | 7 | = | 27 |
| 2 | 8 | 5 | 7 | 1 | 4 | = | 27 |
| 4 | 2 | 8 | 5 | 7 | 1 | = | 27 |
| 5 | 7 | 1 | 4 | 2 | 8 | = | 27 |
| 7 | 1 | 4 | 2 | 8 | 5 | = | 27 |
| 8 | 5 | 7 | 1 | 4 | 2 | = | 27 |
| — | — | — | — | — | — | | |
| 27 | 27 | 27 | 27 | 27 | 27 | | |

And if you can't remember the number there is always an easy way to find it, for 1/7, expressed as a decimal is: ·142857 142857 142857 142857 and so on to infinity.

## *THE NUMBER 526315789473684210*

This deceptively straightforward number is also circular, in the sense that when it is multiplied by any number from 2 to 200 the product can always be read off from a circle made up of the figures of the multiplicand.

If the multiplier is 18 or less the break comes after the digit which is the same as the multiplier, or, where the multiplier has two figures, the same as the last figure of the multiplier. When the cut has been made and the two parts joined add a zero to get the final answer.

As all the digits in the number, except 9 and 0, occur twice you must know which of the pair to make the break after. The rule is this: look at the figures following the two you are choosing between. If the number you are multiplying by is between 2 and

9 make the break before the lower figure, if it lies between 11 and 18 make the break before the higher one. For example:

To multiply 526315789473684210 by 6.

The break must come between 6 and 3 or 6 and 8. The multiplier lies between 2 and 9 so we make the break between 6 and 3.

We write out the answer by taking the figures that follow 6: 315789473684210, carrying on from the beginning of the number and joining on 526, and adding 0 to get the answer:

3157894736842105260

Or to multiply 526315789473684210 by 14.

The break can be made between 4 and 7 or 4 and 2. 14 lies between 11 and 18 so we choose the higher number, and the answer can be read straight off:

7368421052631578940

To multiply by numbers between 19 and 200 is a little more complicated. If they are the numbers we have already dealt with multiplied by 10 – that is 20, 30, 40 and so on up to 180 – there is no problem, for we can carry out the procedure set out above and add 0 to the product. Similarly to multiply by 200 carry on the operation as if multiplying by 2 and add two zeros to the answer.

19, 38, 57, 76 and 95 give products in which all the digits are nines except the last two and (in all cases except 19) the first:

$$\begin{array}{r} 526315789473684210 \\ \times\ 19 \\ \hline 9999999999999999990 \end{array}$$

$$\begin{array}{r} 526315789473684210 \\ \times\ 38 \\ \hline 19999999999999999980 \end{array}$$

$$\begin{array}{r} 526315789473684210 \\ \times\ 57 \\ \hline 29999999999999999970 \end{array}$$

$$\begin{array}{r} 526315789473684210 \\ \times\ 76 \\ \hline 39999999999999999960 \end{array}$$

$$\begin{array}{r} 526315789473684210 \\ \times\ 95 \\ \hline 49999999999999999950 \end{array}$$

These five numbers are special cases. The rule for numbers between 21 and 29 is this: add 1 to the second digit of the multiplier. Multiply the special number by making the break before the lower of the two possible figures. When you reach the last digit reduce it by 1. Insert 1 at the beginning and 0 at the end of the number to arrive at the final product. For example:

$$526315789473684210 \times 27$$

Increase the second digit of the multiplier by 1: $7 + 1 = 8$. Now make a cut between 8 and 4 in our special number and write down the figures

421052631578947368

Reduce the last digit by one. You have:

421052631578947367

Annex a zero to this number and attach the 1 at the beginning to get the product,

14210526315789473670

To multiply the special number by numbers between 31 and 37 use the same method as that set out above, but make the break before the higher of the two possible figures. For example:

$$526315789473684210 \times 34$$

Increase the second digit of the multiplier by 1 to make 5. Make the break between 5 and 7 to get the number 789473684210526315. Reduce the last digit by 1, put a 1 at the beginning and a 0 at the end to arrive at the final product:

17894736842105263140

For multipliers between 39 and 48 the method is the same as for numbers between 21 and 29, but instead of 1 at the beginning of the product write 2, instead of deducting 1 from the last figure of the number deduct 2, and instead of adding 1 to the second digit of the multiplier add 2. For example:

$$526315789473684210 \times 46$$

Add 2 to the second figure of the multiplier: $6 + 2 = 8$. Make a cut at the lower succeeding figure. You get:

421052631578947368

Reduce the last figure by two units. You get:

421052631578947366

Annex a zero to this number and attach the number 2 at the beginning of the number to obtain the product:

242105263157894736600

When the product lies between 49 and 56 the method is the same as for the multipliers between 39 and 48, but the cut is made before the higher of the two possible figures.

When the multiplier is between 58 and 67 the method is as for those between 39 and 48, but where 2 was added in those calculations 3 is added. The cut is made at the lower figure.

For multipliers between 68 and 75 the method is as for 58 to 67, but the break is made at the higher figure.

For multipliers between 77 and 85 insert 4 not 3, and cut at the lower figure and for multipliers between 86 and 94 do the same but cut at the higher one. For multipliers 96 to 104 insert 5 not 4, and cut at the lower figure. For other numbers similar methods apply – you can find them by trial and error, right up to 200. You will find though that 114, 133, 152, 171, and 190 give products made up of a series of nines with two varying digits and a zero or zeros.

## *THE NUMBER 1089*

This, as I mentioned earlier (Chapter 8) has some peculiar traits. For instance look at the pattern that is formed when it is multiplied by the numbers 1 to 9:

$$1089 \times 1 = 1089 \qquad 9801 = 1089 \times 9$$
$$1089 \times 2 = 2178 \qquad 8712 = 1089 \times 8$$
$$1089 \times 3 = 3267 \qquad 7623 = 1089 \times 7$$
$$1089 \times 4 = 4356 \qquad 6534 = 1089 \times 6$$
$$1089 \times 5 = 5445$$

Or try this trick: write 1089 on a piece of paper and put it in your pocket. Now ask someone to think of a three-digit number, the first and last digits of which differ by at least 2. Let us suppose he chose 517. Now ask him to reverse the digits and subtract the higher number from the lesser ($715 - 517 = 198$). Finally ask him to add this number to itself reversed ($198 + 891 = 1089$). No matter what number he starts with you will always have the final answer – 1089 – in your pocket!

## *THE NUMBER 12345679*

This number too appeared earlier (Chapter 2). Now we will look at some of its other peculiarities. Multiplying it by 3, and multiples of 3, gives a curious set of patterns:

$$\begin{aligned}
12345679 \times 3 &= 37037037 \\
12345679 \times 6 &= 74074074 \\
12345679 \times 9 &= 111111111 \\
12345679 \times 12 &= 148148148 \\
12345679 \times 15 &= 185185185 \\
12345679 \times 18 &= 222222222
\end{aligned}$$

and so on.

Multiplying by 999999999 gives a number which is a mirror image of itself:

$$12345679 \times 999999999 = 12345678987654321$$

Incidentally, this product is a perfect square:

$$111111111^2 = 12345678987654321$$

## *PRIME NUMBERS*

A positive integer, a whole number, greater than one, that cannot be expressed as a product of two positive integers, neither of which is 1, is known as a prime number. Any positive integer greater than 1 that is *not* a prime number can be expressed as a product of two or more prime numbers. For example, $4 = 2 \times 2$; $6 = 3 \times 2$; $8 = 4 \times 2$ and so on.

The first few prime numbers 2, 3, 5, 7, 11 etc, are easy to spot. Until 1952 the highest prime number known was $2^{127} - 1$, which written out in full is

170141183460469231731687303715884105727

Then five higher primes were found by computer the largest being $2^{2281} - 1$. More recently a computer in Sweden has been used to prove that $2^{3271} - 1$ is also a prime number.

Two numbers that were thought to be primes on the other hand have lately been shown to have factors. They are

1757051 ($1291 \times 1361$) and 222221 ($619 \times 359$)

## *STRANGE ADDITION*

| | | | | | | | | |
|---|---|---|---|---|---|---|---|---|
| 1 | | | | | | | | |
| 2 | 6 | | | | | | | |
| 3 | 9 | 15 | | | | | | |
| 4 | 12 | 20 | 28 | | | | | |
| 5 | 15 | 25 | 35 | 45 | | | | |
| 6 | 18 | 30 | 42 | 54 | 66 | | | |
| 7 | 21 | 35 | 49 | 63 | 77 | 91 | | |
| 8 | 24 | 40 | 56 | 72 | 88 | 104 | 120 | |
| 9 | 27 | 45 | 63 | 81 | 99 | 117 | 135 | 153 |

In the table above the horizontal lines are arithmetical progressions. The difference between each number and the one to its right is twice the figure that stands at the beginning of the row. (Row 7 for example can be worked out by adding 14, first to 7 and then to each successive total.)

But how would you find the sum of all the numbers in any row? There is no need to add them – it is the same as the cube of the number which stands at the beginning of the row. For instance the total of the numbers in row 6 is $6 \times 6 \times 6 = 216$.

## *AND SOME STRANGE FRACTIONS*

Here are two fractions, with one as numerator and with two different uneven multiples of 9 as denominator:

$$\frac{1}{9 \times 3} = \frac{1}{27} \quad \text{and} \quad \frac{1}{9 \times 17} = \frac{1}{153}$$

Now let us take the first fraction and reduce it to decimals, by dividing the numerator by the denominator:

$$1/27 = \cdot037037037037 \quad \text{to infinity}$$

and take just two significant figures of the first three places of decimals: 37. Now watch the pattern this number forms when multiplied by 3 and its multiples:

$$\begin{aligned}
37 \times 3 &= 111 \\
37 \times 6 &= 222 \\
37 \times 9 &= 333 \\
37 \times 12 &= 444 \\
37 \times 15 &= 555 \\
37 \times 18 &= 666 \\
37 \times 21 &= 777 \\
37 \times 24 &= 888 \\
37 \times 27 &= 999
\end{aligned}$$

Now take the second fraction and reduce it to decimals:

$$1/153 = \cdot 0065359477124183$$

Take all the significant figures, multiply by 17, and multiples of 17 and watch the pattern formed:

$$65359477124183 \times 17 = 1111111111111111$$
$$65359477124183 \times 34 = 2222222222222222$$
$$65359477124183 \times 51 = 3333333333333333$$
$$65359477124183 \times 68 = 4444444444444444$$
$$65359477124183 \times 85 = 5555555555555555$$
$$65359477124183 \times 102 = 6666666666666666$$
$$65359477124183 \times 119 = 7777777777777777$$
$$65359477124183 \times 136 = 8888888888888888$$
$$65359477124183 \times 153 = 9999999999999999$$

## *REVERSALS*

| | |
|---|---|
| $9 + 9 = 18$ | $81 = 9 \times 9$ |
| $24 + 3 = 27$ | $72 = 24 \times 3$ |
| $47 + 2 = 49$ | $94 = 47 \times 2$ |
| $497 + 2 = 499$ | $994 = 497 \times 2$ |

## *CASTING OUT THE NINES*

In other chapters I have described ways of checking answers by casting out the nines. Casting out the nines – by repeatedly subtracting 9 until a remainder of less than 9 is left, or, which amounts to the same thing, dividing by 9 and noting the remainder – can be done in an oddly simple way. The remainder when a number has been divided by 9 is the same as the sum of the digits (or, when that sum gives a number with two digits the sum of those digits). As the remainder – not the number of nines – is what you are after you can arrive at it directly. Here are two examples:

Cast the nines from 67 and find the remainder.

| | | |
|---|---|---|
| 67 | 9)67(7 | $6 + 7 = 13$ |
| −9 | 63 | $3 + 1 = 4$ |
| — | — | |
| 58 | 4 | |
| −9 | | |
| — | | |
| 49 | | |
| −9 | | |
| — | | |
| 40 | | |
| −9 | | |
| — | | |
| 31 | | |
| −9 | | |
| — | | |
| 22 | | |
| −9 | | |
| — | | |
| 13 | | |
| −9 | | |
| — | | |
| 4 | | |

Cast out the nines from 44 and find the remainder.

```
 44        9)44(4        4 + 4 = 8
 -9          36
 --          --
 35           8
 -9
 --
 26
 -9
 --
 17
 -9
 --
  8
```

## *PI*

Pi is the number which expresses the ratio of the circumference of a circle to its diameter. While it has been proved (by Lindemann, in 1882) that it cannot be a repeating decimal (in mathematical terms that it is a transcendental number and not the root of a finite algebraic equation with integral coefficients), it has been calculated to a degree of accuracy that would once have seemed impossible. The story of the accuracy to which the value of Pi is known is an interesting one.

In the Bible, the value of Pi given is 3. Archimedes declared the value of Pi to be less than 3 1/7 but greater than 3 10/71. The value generally used today, 3·1416, was known at the time of Ptolemy (A.D. 150). In the 16th century F. Vieta calculated Pi to ten places of decimals. At the end of the 16th century, Ludolph van Ceulen calculated thirty-five decimal places. In his will he requested that these thirty-five numbers be engraved on his tombstone. This was done, and this number is still known in Germany as the 'Ludolphian' number, in his honour.

Now computers have come to the aid of the mathematicians. In 1949 a group using the rather primitive computer Eniac took 70 hours to calculate Pi to 2037 decimal places, while recently Daniel Shanks and John Wrench Jnr have published Pi to 100000 places, arrived at in 8 hours and 43 minutes, using an IBM 7090 system. They used the following formula:

$$\pi = 24 \tan^{-1}\frac{1}{8} + 8 \tan^{-1}\frac{1}{57} + 4 \tan^{-1}\frac{1}{238}$$

It took them 3 hours and 7 minutes to compute: $8 \tan^{-1} 1/57$; 2 hours and 20 minutes to compute: $4 \tan^{-1} 1/238$; 2 hours and 34 minutes to compute: $24 \tan^{-1} 1/8$; and 42 minutes for conversion (binary to decimal).

Mnemonics are often used to help remember the digits of Pi. In the sentence '*May I have a large container of coffee?*' the value of Pi to seven places of decimals is contained. The number of letters in each word corresponds to the successive integers in the decimal expansion of Pi.

Sir James Jeans came out with the following sentence, in which the value of Pi is contained to fourteen decimal places:

'*How I want a drink, alcoholic of course, after the heavy chapters involving quantum mechanics.*'
3 14 15 92 65 35 89 79

Adam C. Orn of Chicago published in the literary digest of Chicago of 20 January 1906, the following poem that contains Pi to thirty decimal places:

Now I – even I, would celebrate
In Rhymes unapt the great
Immortal Syracusan rivaled never more,

Who in his wonderous lore
Passed on before,
Left men his guidance
How to circles mensurate.

31415926535897543238452643383279

It is also said that this French poem contains Pi to thirty places of decimals:

Que j'aime à faire apprendre
Un nombre utile aux sages
Immortal Archimède Artiste Engènieur
Qui de ton jugement peut priser la valeur
Pour moi ton problème
A les pareils avantages.

However, there are two mistakes in the last line. Where 'A les' represents the digits 1 and 3, the correct digits are 3 and 2.

# 15

# *TRICKS and PUZZLES*

Creative mathematics is divided from play only very vaguely. It is said that Leibnitz spent a considerable amount of time studying solitaire, and Einstein's bookshelves were stacked with books on mathematical games, so this section of tricks and puzzles may lead you to think more like a mathematician than some of the methods for solving standard problems in earlier chapters. Much of science is after all finding mathematical solutions or descriptions for puzzles posed by nature.

### *Who Sits by Who*

It was a boarding school. In the hostel dining-room the tables seated ten and on the first day of school, ten boys stood around one table and argued about who should sit next to whom.

One of the boys suggested that they should sit according to the alphabetical order of their names. Another proposed that the seating should be planned according to their heights. And yet another suggested that they should have a raffle and pick out the names against the numbers of the seats. The argument continued.

Then the head boy of the class came out with a splendid proposal: 'Let us all sit today according to the position we are standing around the table. Tomorrow we will change the position. The day after tomorrow we again change the position. This way each day we will sit in a different order. And when we are back in the same position as we are sitting in today, we shall celebrate it with a special dinner, and I shall stand the expenses.' This proposition seemed very reasonable to all of them and they agreed.

But the day never came when the head boy had to throw the

special dinner party, even when they had all graduated and were ready to leave school! There are too many different ways 10 boys could sit around a table. The possible number of combinations is given by the calculation:

$$1 \times 2 \times 3 \times 4 \times 5 \times 6 \times 7 \times 8 \times 9 \times 10 = 3628800$$

If instead of 10 there were 13 boys, the possible combinations would have been:

$$1 \times 2 \times 3 \times 4 \times 5 \times 6 \times 7 \times 8 \times 9 \times 10 \times 11 \times 12 \times 13 = 6227020800$$

If there were only two boys, the combination could have been just: $1 \times 2 = 2$.

If there were 3 boys: $1 \times 2 \times 3 = 6$.

If there were 4 boys: $1 \times 2 \times 3 \times 4 = 24$.

If there were 5 boys: $1 \times 2 \times 3 \times 4 \times 5 = 120$.

And if there were 15 boys, the maximum number of combinations would be: $1 \times 2 \times 3 \times 4 \times 5 \times 6 \times 7 \times 8 \times 9 \times 10 \times 11 \times 12 \times 13 \times 14 \times 15 = 1307674365000$.

If you consider 25 boys, the number of combinations would be: $1 \times 2 \times 3 \times 4 \times 5 \times 6 \times 7 \times 8 \times 9 \times 10 \times 11 \times 12 \times 13 \times 14 \times 15 \times 16 \times 17 \times 18 \times 19 \times 20 \times 21 \times 22 \times 23 \times 24 \times 25 = 15511210043330985984000000$.

It would be eternity before a party of 25 people could exhaust all the combinations.

### *The Boy who could only Add*

Shambhu came to the city from a remote village. Though he had never attended school he had been taught by his father how to add and divide by two. This was all the arithmetic he had, yet, he told a friend, he could rapidly multiply numbers with two digits. The friend was surprised.

'Very well,' he said, 'what is 43 times 87?'

Shambhu came up with the answer, 3741, in a few minutes. This is how he did it.

He took a piece of paper and wrote the number 43 on the left and 87 on the right. He divided 43 by 2 and divided the answer by 2 again. He continued dividing by 2 until he reached 1, ignoring the remainders at each stage except the last. Then he doubled the 87 successively until he had an equal number of figures in each column.

Having written these two columns of numbers he cancelled every number in the right-hand column that was opposite an even number in the left-hand column. Then he added up the numbers remaining in the right-hand column to get the answer. This is how his paper looked:

| 43 | 87 | 87 |
|---|---|---|
| 21 | 174 | 174 |
| 10 | ~~348~~ | 696 |
| 5 | 696 | 2784 |
| 2 | ~~1392~~ | 3741 |
| 1 | 2784 | |

*Dividing the Elephants*

The Maharaja of an Indian State invited three artists to perform at his court. He was much impressed, and honoured them with gold and gifts.

On the day of their departure he presented the three of them with seventeen elephants, expressing a wish that the oldest of the artists should get one-half of the number of elephants, the middle one one-third of the number, and the last one one-ninth.

When the artists returned home, they were absolutely puzzled: how could they divide the elephants according to the wishes of the Maharaja? One-half of the seventeen elephants meant that the oldest artist gets eight and a half elephants. What good is an elephant cut in two! They argued, and the leading citizens of the town were called to settle the dispute. They tried in vain to carry out the will of the Maharaja for they found themselves constantly held up, by the impossibility of making any share a whole number of elephants. They were about to give up in despair when someone in the crowd suggested that the temple priest, a wise, very learned man, be called in to settle the affair.

The temple priest listened to the whole story in detail, got up from his seat and asked to be excused. Then he went back to the temple and returned in a few minutes with the temple elephant. 'Now,' he remarked, 'the problem can be easily solved. I am temporarily gifting this temple elephant, and pooling it in with the other elephants.'

There were altogether eighteen elephants, with the temple elephant thrown in.

The priest addressed the oldest artist. 'Here, you take away half the number of elephants,' he said and presented him with nine elephants.

Then he looked at the middle artist and remarked, 'You are entitled to one-third of the number, and here are your six elephants.' He gave away six elephants to him.

Now there were three elephants remaining including the temple elephant.

Then to the youngest artist he said, 'Your share is one-ninth of the total number of elephants. Therefore you get two elephants.' He gave away two elephants to him.

The remaining elephant was the temple elephant. 'I take back the temporary gift I had made.' So saying he walked away with his elephant, back to the temple. The will of the Maharaja was followed to the letter.

### *The Lily Pond*

In India water lilies grow extremely rapidly. In one pond a lily grew so fast that each day it covered a surface double that which it covered the day before. At the end of the thirtieth day it entirely covered the pond, in which it grew. But how long would it take two water lilies of the same size at the outset and at the same rate of growth to cover the same pond?

Naturally, one is bound to think that if it takes one water lily 30 days to cover the pond, it should take two water lilies 15 days to do the same thing – that is, half the time. But this answer is wrong. Since one plant doubles its surface every day, it will have half the surface left at the end of the 29th day. Therefore with two water lilies at the outset, each will cover its half of the pond at the end of the 29th day, and not before.

### *Which Vacation?*

It was vacation time. Two men were travelling in a train. The older man asked the young man seated opposite to him how old he was. The young man replied, 'The day before yesterday I was 19, and next year I will be 22.' Can you guess during which vacation and on what day of the month these men were travelling?

It must have been the Christmas vacation, and on 1 January, his birthday being on 31 December. On 30 December he was still 19. By 1 January he was 20. And on 31 December in the same year he would be 21 and the next year on 31 December he would be 22.

## *As Old as the Year of Your Birth*

The grandson told his grandfather that in 1932 he was as old as the last two digits of his birth year and half as old as the century in which he was born. The grandfather replied that the same applied to him too. That's quite impossible said the grandson. But the grandfather went ahead and proved his statement. Can you guess how old the grandson and the grandfather were?

The grandson was born in the twentieth century and therefore the first two digits of his birth year are 19. The other two digits are half of 32, 16. The grandson was born in 1916.

The grandfather was born in the nineteenth century. The first two digits of his birth year are 18. The rest of the digits doubled must be equal to 132. Therefore the number we want is half of 132. And that number is 66. The grandfather was born in 1866. In 1932 he was 66.

## *Broken Stones*

A store-keeper had a stone weighing 40 pounds which used to weigh goods in 40-pound lots. But one day it fell and broke into four pieces. He was about to dump them in the dustbin when a neighbouring store-keeper stopped him and showed him five pieces of stone he himself had, which were the broken pieces of a stone that used to weigh 31 pounds. He explained that with these five pieces he could weigh any whole number of pounds from 1 to 31, and advised him to check the pieces before throwing them away.

The store-keeper found to his surprise that with the four stones he could weigh any whole number of pounds from 1 to 40.

What are the weights of the pieces of stone used by the store-keeper and his neighbour?

The store-keeper's 40-pound stone broke into pieces weighing 1, 3, 9 and 27 pounds. His neighbour's 31-pound stone broke into pieces weighing 1, 2, 4, 8 and 16 pounds.

*Spreading Rumours*

It is amazing how fast rumours can spread. Things witnessed by just one or two people became common knowledge within a couple of hours. The speed with which rumours travel is extraordinary. It is fantastic! But when we work the thing out mathematically, there seems to be nothing fantastic about it. Let us examine a case and see how it works out.

Manju Guha comes from the city to visit her relatives living in a small town. She arrives in their home in the morning at seven o'clock. She has a very interesting piece of news from the city which she conveys to the family with whom she is staying. There are 4 members in the family and it takes Manju Guha exactly 10 minutes to tell them the gossip. Now can you guess how long it would take for this news to spread around all the 87,300 inhabitants of the town, if each person took just 10 minutes to tell it to another 4 people?

| | |
|---|---|
| Manju Guha arrives at 7 am | |
| At 7 am the news is known to | 1 |
| It takes her 10 minutes to convey the news to 4 | |
| By 7.10 am the matter is known to (1 + 4) | 5 |
| Each of the four hastens to tell another. | |
| So by 7.20 am the gossip is known to 5 + (4 × 4) | 21 |
| By 7.30 am the gossip is known to 21 + (16 × 4) | 85 |
| By 7.40 am the gossip spreads to 85 + (64 × 4) | 341 |
| By 7.50 am to 341 + (256 × 4) | 1365 |
| By 8 am to 1365 + (1024 × 4) | 5461 |
| By now the rumour gathers momentum and | |

By 8.10 am is known to $5461 + (4096 \times 4)$ 21845
By 8.20 am the gossip spreads to $21845 + (16384 \times 4)$ 87381

The news that was known to only one person at 7 am is known by everyone in town by 8.20 am, 1 hour and 20 minutes later. Mathematically the whole calculation can be condensed into the following addition:

$$1 + 4 + (4 \times 4) + (4 \times 4 \times 4) + (4 \times 4 \times 4 \times 4) + (4 \times 4 \times 4 \times 4 \times 4) + (4 \times 4 \times 4 \times 4 \times 4 \times 4) + (4 \times 4 \times 4 \times 4 \times 4 \times 4 \times 4) + (4 \times 4 \times 4 \times 4 \times 4 \times 4 \times 4 \times 4)$$

that is: $1 + 4 + 4^2 + 4^3 + 4^4 + 4^5 + 4^6 + 4^7 + 4^8$
which equals:

$$\begin{array}{r} 1 \\ 4 \\ 16 \\ 64 \\ 256 \\ 1024 \\ 4096 \\ 16384 \\ 65536 \\ \hline 87381 \end{array}$$

The people of this particular town, not being gossip-mongering types, conveyed the rumour to only 4 people each. But if each person had shared the rumour with more people, it would have spread with even greater speed!

### *How Many Tickets?*

Imagine a small railway line with 30 railway stations. Can you guess how many different kinds of tickets the railway has to have printed?

At each of the 30 stations the passengers can get tickets for any of the other 29 stations. Therefore the number of tickets is:

$$30 \times 29 = 870$$

### *Proving the Impossible*

Can you prove $1 = 2$?

Impossible you think? All you need is a very elementary application of algebra:

Let

$$a = b$$

Then

$$a \times b = b \times b$$

$$ab = b^2$$

Subtract $a^2$ from either side:

$$ab - a^2 = b^2 - a^2$$

or

$$a(b - a) = (b + a)(b - a)$$

Divide both sides by $b - a$.

Then

$$a = b + a$$

If $a = 1$,

$$1 = 1 + 1$$

Therefore

$$1 = 2$$

### *All Fours*

Can you combine four 4s with any other mathematical symbol except numbers and produce all the whole numbers from 1 to 18? You may use the four symbols of addition, subtraction, multiplication and division, the square root symbol, and the decimal point. You may, if you wish, put the fours side by side.

$$\frac{4+4}{4+4}=1$$

$$\frac{4\times 4}{4+4}=2$$

$$\frac{4+4+4}{4}=3$$

$$\frac{4+4}{4}+\sqrt{4}=4$$

$$\frac{(4\times 4)+4}{4}=5$$

$$\frac{4+4+4}{\sqrt{4}}=6$$

$$\frac{44}{4}-4=7$$

$$(4\times 4)-(4+4)=8$$

$$4+4+\frac{4}{4}=9$$

$$\frac{44-4}{4}=10$$

$$\frac{4}{.4}+\frac{4}{4}=11$$

$$\frac{44+4}{4}=12$$

$$\frac{44}{4}+\sqrt{4}=13$$

$$4+4+4+\sqrt{4}=14$$

$$\frac{44}{4}+4=15$$

$$4+4+4+4=16$$

$$(4\times 4)+\frac{4}{4}=17$$

$$4\times 4+\frac{4}{\sqrt{4}}=18$$

*Finding the Counterfeit*

The gold merchant examined the nine gold coins he had purchased at an auction that morning. He had been told by the auctioneer that one of the coins was counterfeit, and that the counterfeit coin weighed a gram less than the genuine ones. He was baffled as to how to pick it out.

His old clerk came to his assistance. 'In just two weighings I can spot it,' he said. And he did just that. How did he do it?

He split the nine coins in three groups of three coins each. At the first weighing he took two groups and weighed them against each other. If they balanced, it meant that the third group contained the counterfeit coin. If they did not, whichever group weighed lighter contained the bad coin. Now he knew which group contained the counterfeit coin. In the next weighing, he took two of the three coins from the group containing the counterfeit coin and weighed them against each other. If they balanced the third coin was counterfeit. Otherwise the lighter one in the balance was counterfeit.

### *Transformations*

Can you write 23 with only twos, 45 with fours and 1000 with only nines?

Yes!

$$22 + 2/2 \qquad 44 + 4/4 \qquad 999 + 9/9$$

### *The Millionaire and the Merchant*

It was a beautiful afternoon, the 1st of June. A millionaire walked into a pub to have a drink. He was joined by a local merchant.

In the course of conversation, the merchant made a proposition. He offered to pay the millionaire $1000 a day for the rest of the month, starting from that very day, and in return he expected only pennies. Just one cent on that first day, double that, 2 cents, the next day, 4 cents on the third day, and so on, doubling the amount every day for the rest of the month.

The millionaire burst into loud laughter, thinking the merchant a fool, and agreed. They shook hands and the deal was closed.

The merchant took out a bundle of $1000 and handed it over to the millionaire. The millionaire gave the merchant a penny in return.

It was agreed that they would meet every day in the same place and at the same time for the rest of the month, to carry out their bargain. So the next day the millionaire cheerfully took the bundle of $1000 from the merchant and handed him over 2 cents. And the next day again the merchant handed over the bundle of $1000 to the mocking millionaire, and in return got his 4 cents. In the first ten days the millionaire received $10000 from the merchant and in return had paid him the mere pittance of $10·23.

But by the 17th day, the millionaire felt a little uncomfortable, when he had to pay 65,536 cents, that is $655·36 to the merchant in return for the $1000 he got from him, for he saw that next day he must pay more than he received: 131072 cents, that is $1310·72.

And when he looked to the future he got very worried indeed. For day after day, as the merchant handed over his $1000 he got back from the millionaire amounts many times more than that.

On the last day of the deal, the 30th day, the millionaire had to liquidate his assets to the last cent in order to honour his agreement with the merchant.

Can you guess the amount of loss suffered by the millionaire?

On the 1st day the millionaire pays 1 cent
On the 2nd day he has paid $1 + 2 = 3$
On the 3rd day he has paid $3 + 4 = 7$
On the 4th day he has paid $7 + 8 = 15$
On the 5th day he has paid $15 + 16 = 31$
On the 6th day he has paid $31 + 32 = 63$
On the 7th day he has paid $63 + 64 = 127$

The whole calculation can be represented as:

$$1 + 2 + 2^2 + 2^3 + 2^4 + 2^5 + 2^6 = 127$$

On the 8th day it is:

$$1 + 2 + 2^2 + 2^3 + 2^4 + 2^5 + 2^6 + 2^7 = 255$$

and so on.

On the 18th day he begins to pay more than he receives and on the last day, the 30th of June, the millionaire has to pay a sum of \$5368709·12. The total amount paid by the millionaire to the merchant in the 30 days works out to

$$(2^{30} - 1)\text{cent} = 1073741823 \text{ cents} = \$10737418{\cdot}23$$

whereas he had received from the merchant in the 30 days \$30000. Therefore the loss suffered by the millionaire is equivalent to:

$$\$10737418{\cdot}23 - \$30000 = \$10707418{\cdot}23$$

*Thousands by Eights*

Can you combine eight 8s with any other mathematical symbols except numbers to represent exactly 1000? You may use the plus, minus, division and multiplication signs.

In an earlier chapter of this book, I gave one solution to this problem, but there are many more both simpler and more complex. Here are some of them:

$$\frac{8+8}{8}(8 \times 8 \times 8 - 8) - 8 = 1000$$

$$\frac{8888 - 888}{8} = 1000$$

$$\left(\frac{88-8}{.8}\right)\left(8+\frac{8+8}{8}\right)=1000$$

$$8\Big((8\times 8)+(8\times 8)\Big)-8-8-8=1000$$

$$\left(8\,(8+8)-\frac{8+8+8}{8}\right)8=1000$$

*Telling and Not Telling*

A girl, asked how old she was, did not wish to give out her age directly, and yet did not want to be rude and refuse to give it all. Therefore she said, 'My age three years hence multiplied by three, less three times my age three years ago, will give you my present age.' Can you guess her age?

Let us assume her present age is X years. Three years hence her age will be X + 3 years. Her age three years ago was X − 3 years.

$$3(X+3)-3(X-3)=X$$

Therefore, X = 18, which is her age. Three years hence she will be 21 and three years back she was 15.

$$(3\times 21)-(3\times 15)=63-45=18$$

*Guessing Birthdays*

You can tell a friend the month and the date of his birth very easily.

First of all, ask him to keep in mind two numbers, the number of the month on which he was born and the number of the date of the month. (The months, of course, are numbered 1 to 12, from January to December.) Then you ask him to multiply the number of the month by 5, add to this 6, and multiply it by 4, and then add 9. Once again, ask him to multiply the number, this time by 5, and add to it the number of the date on which he was born.

When he finishes the calculations, ask him for the final result and mentally subtract 165 from it. The remainder is the answer. The last two digits of the number, give you the date of the month and the first digit or the first two digits give the number of the month.

For example, if the final result is 1269, when you subtract 165 you get 1104. From this number you know that he was born on 4 November. The trick behind it is, of course, the 165.

The directions you give your friend are a disguised way of adding 165 to the number of the month, multiplied by one hundred. When the number 165 is taken away from the total, the number of the day, and one hundred times the number of the month are left!

### *False Promises*

Some years back the following advertisement appeared in the popular press:

UNIQUE OPPORTUNITY! UNIQUE OPPORTUNITY!
PRESSURE COOKER ONLY $1

DELAY MEANS DISAPPOINTMENT
RUSH!!!

AND WRITE FOR THE FREE APPLICATION FORM
OFFER GOOD ONLY WHILE STOCKS LAST

Housewives did indeed rush to write for free application forms. Pressure cookers, in those days, were usually sold for $5. In return for her letter, what each housewife received however was not an application form – but four coupons which she was told to sell to her friends at $1 each. She was then to send the $4 thus collected, along with her own $1 to the company. On receipt of this amount, the company sent her a pressure cooker. And many housewives did get their pressure cookers for $1 – the other $4 came from the purses of their friends.

At first, every housewife got her pressure cooker for only $1 and there seemed to be nothing fraudulent about the transaction. The company kept its promise, and gave the housewives pressure cookers for $1, and yet was not losing a penny in the bargain. They got their full price for the goods.

However, it was obvious that the whole thing was a kind of a swindle. People, somewhere, must have been losing money. Sooner than expected came the time when coupon holders found it impossible to dispose of the extra four coupons they had to sell if they were to claim their cooker. The first group of housewives, who got their coupons direct from the company, found no difficulty in finding buyers for them. Each of these housewives was able to draw four new partners into the deal. And each of these four new partners had had to dispose of their coupons to four other partners.

Let us calculate and see how fast the number of coupon holders increased:

| | |
|---|---|
| *first round* | 1 |
| *second round* | 4 |
| *third round* | 20 |
| *fourth round* | 100 |
| *fifth round* | 500 |
| *sixth round* | 2500 |

| | |
|---|---|
| *seventh round* | 12500 |
| *eighth round* | 62500 |
| *ninth round* | 312500 |
| *tenth round* | 1562500 |
| *total* | 1953125 |

If everyone in a town of two million joined all would be covered by the tenth round. But only one-fifth of them would get their pressure cookers. The rest being left with valueless coupons.

*Missing Numbers*

Here is a multiplication with more than half of the digits expressed by X's. Can you spot the missing digits?

```
    X1X
    3X2
    ---
    X3X
   3X2X
  X2X5
  -----
  1X8X30
```

The last digit in the product is a zero. Therefore the digit in the third line has got to be a zero. The X at the end of the first line must be a number that gives a number ending with zero if multiplied by 2, and with 5 if multiplied by 3, as the number in the fifth line ends with a 5. Only 5 will do this.

The X in the second line must be 8 – that is the only number which, when multiplied by 15, gives a number ending with 20 in the fourth line. It is clear that the first X in line 1 is a 4, because only 4 multiplied by 8 gives a number that begins with 3 in the fourth line. Now there will be no difficulty in spotting the remaining digits:

```
   415
   382
   ---
   830
  3320
 1245
------
158530
```

*Thirty Rewritten*

Can you write 30, using any three identical digits except 5s? Here are three solutions:

$$6 \times 6 - 6 = 30; \; 3^3 + 3 = 30; \; 33 - 3 = 30$$

*Dividing the Cake*

Three men were travelling in a railway compartment. One of them was a very wealthy businessman. The other two had food with them and invited the businessman to join them in their meal. One man had brought 5 chapathis and the other had 3. The three of them ate together and the food was shared equally.

On parting, the businessman gave 8 gold coins to the two men, as a token of his appreciation of their hospitality, and told them to share the coins in proportion to their contribution of food.

An argument started. The man who had brought 5 chapathis felt that he should get 6 coins and the other man should get 2 coins. But the man who had brought 3 chapathis felt that he should get 3 coins and the other man should get 5 coins.

As the argument continued a crowd gathered. Then a wise old man, who was listening, intervened and offered to settle the dispute. He heard both sides of the case and declared that the man who had brought 5 chapathis should get 7 coins and the man who had brought 3 chapathis should get only one. The people in the crowd protested and demanded an explanation. His explanation was this: the food was shared equally by three men. Each chapathi was portioned into three equal parts. Therefore the 8 chapathis together were portioned into 24 equal pieces. Of the 24 pieces, each man ate 8 pieces. The man who had brought 5 chapathis contributed 15 pieces, and the man who had brought 3 chapathis contributed 9 pieces. Out of the 15 pieces, the man who had brought 5 chapathis himself ate 8 pieces and contributed to the guest 7 pieces. And out of the 9 pieces, the man who had brought 3 chapathis, himself ate 8 pieces and contributed to the guest only 1 piece.

So the man with the 5 chapathis gets 7 coins and the man with 3 chapathis gets only 1.

### *All the Digits*

Do you notice anything interesting in the following multiplication?

$$138 \times 42 = 5796$$

There are nine digits and all are different. Can you think of other such combinations?

Here is a group of nine:

$$12 \times 483 = 5796$$
$$18 \times 297 = 5346$$
$$39 \times 186 = 7254$$
$$48 \times 159 = 7632$$
$$27 \times 198 = 5346$$
$$28 \times 157 = 4396$$
$$4 \times 1738 = 6952$$
$$4 \times 1963 = 7852$$

## *Another Impossible Proof*

Can you prove $45 - 45 = 45$?

Write down the figures from 9 to 1

| | | |
|---|---|---|
| | 987654321 | add up the digits $9+8+7+6+5+4+3+2+1=45$ |
| Subtract | 123456789 | Add up the digits $1+2+3+4+5+6+7+8+9=45$ |
| Result | 864197532 | Add up the digits $8+6+4+1+9+7+5+3+2=45$ |

Therefore $45 - 45 = 45$

## *Dwindling Gifts*

Two fathers gave their sons some money. One father gave his son \$300 and the other son gave his son \$150. Then the two sons counted up their money, they found that between them they had only \$300. How do you explain this?

While there are two fathers and two sons, there are only three people – grandfather, father and son. The grandfather gave his son \$300. Out of this \$300 the father gave his son \$150. Thus between the father and the son – two sons – they have only \$300.

### *Making the Signs Right*

$$3328 = 18$$

This looks absurd, but it can be made sense of by the insertion of mathematical signs, without altering any of the figures, or the position of the equals sign. Here is the solution:

$$\sqrt{332 - 8} = 18$$

### *Five Digits*

Can you arrange the digits 1, 2, 3, 4 and 5 in such a way that, with the aid of simple mathematical signs, they will make 111, 222, 333 and 999? Here is the solution:

$$135 - 24 = 111$$
$$214 + 5 + 3 = 222$$
$$345 - 12 = 333$$
$$(5^3 \times 4 \times 2) - 1 = 999$$

### *Making One Hundred*

Can you, with only the digits 1, 2, 3, 4, 5, 6, 7 and 8 and the addition and subtraction signs make 100?

$$86 + 2 + 4 + 5 + 7 - 1 - 3 = 100$$

### *Smallest Integer*

Can you spot the smallest integer that can be written with two digits?

$$\frac{1}{1}$$

*One from Ten*

Can you write 1, by using all the ten digits?

$$\frac{148}{296} + \frac{35}{70} = 1$$

*All Nines*

Using only five nines, can you write 10?

There are several ways of doing this:

$$9 + \frac{99}{99} = 10 \qquad \frac{99}{9} - \frac{9}{9} = 10$$

$$\left(9 + \frac{9}{9}\right)^{9/9} = 10$$

*One Hundred from Ten*

Can you write 100 by using all the ten digits? Give four different solutions.

$$70 + 24\frac{9}{18} + 5\frac{3}{6} = 100$$

$$50\frac{1}{2} + 49\frac{38}{76} = 100$$

$$87 + 9\frac{4}{5} + 3\frac{12}{60} = 100$$

$$80\frac{27}{54} + 19\frac{3}{6} = 100$$

*Magic and Thirty-seven*

Write 37 on a piece of paper and keep it in your pocket. Ask someone to suggest any number formed of three identical figures. Ask him to add up these three figures and divide the original number by the answer.

Whatever number he chose you will always have the answer – 37 – on the paper in your pocket.

555 divided by $5 + 5 + 5$ gives 37
999 divided by $9 + 9 + 9$ gives 37
888 divided by $8 + 8 + 8$ gives 37

and so on.

*Hare and Tortoise*

Two cars do the journey between cities 200 miles apart. One car at 50 miles an hour one way and 40 miles an hour on the return journey. The other car at 45 miles an hour on both journeys. Which of the two cars covers the distance in less time?

The speed of the second car is the average between the speed of the first and second journeys of the first car. But they do not cover the distance in the same time.

The first car takes, on its first journey, $200 \div 50 = 4$ hours, and on its second journey $200 \div 40 = 5$ hours. A total for both journeys of 9 hours. The second car, on both journeys, takes $400 \div 45 = 8$ hours and 53 minutes.

The second car covers the distance in a shorter time.

*All Threes*

Can you take four 3s and simple mathematical signs and arrange them so that they will make in turn 11, 37, 80 and 100?

$$\frac{33}{\sqrt{3} \times \sqrt{3}} = 11$$

$$3^3 + \frac{3}{\cdot 3} = 37$$

$$\frac{3^3 - 3}{\cdot 3} = 80$$

$$\frac{3}{\cdot 3} \times \frac{3}{\cdot 3} = 100$$

### *The Chess Board*

Perhaps the most famous of the stories about people who were caught out by the way repeated doubling quickly leads to huge totals is that of King Purushottama and the learned Pandit.

The king, proud of his abilities at chess, had challenged and beaten all comers until, after a long struggle, the Pandit beat him.

The apparently insignificant reward he asked was 1 grain of rice for the first square of the chess board, 2 for the second, 4 for the third, 8 for the fourth and so on.

When the sack of rice was slow in coming the king called for his ministers and asked why.

The total number of grains – 18446744073709551615 – was not only beyond the means of his granaries, but beyond the means of the granaries of all the world. The formula for arriving at this number – $2^{64} - 1$ – turns up in another famous legend....

### *The Tower of Brahma*

This, so the legend has it, is a temple in Benaras. They say that when the great lord Brahma created the world he put up three diamond sticks, mounted on a brass plate, beneath the dome that marks the centre of the world. Upon one of the sticks he placed 64 gold discs. The biggest at the bottom, all decreasing in size, with the smallest at the top. The temple priests have the tasks of transposing the discs from one stick to the other, using the third as an aid. They work day and night, but must transpose only one disc at a time and must not put a bigger disc on top of a smaller one. When their task is complete, the legend says, the world will disappear in a clap of thunder.

Now the total number of transpositions needed to shift the 64 discs is, again, $2^{64} - 1$, or 18446744073709551615. If every transposition takes a second it will still take 500000 million years to get the job done.

### *1729*

This is known as Ramanujam's number. When the great Indian mathematician was lying ill in hospital, Professor Hardy, who had been instrumental in bringing him to Cambridge, came to visit him. The taxi I came in, he said to Ramanujam, had a very boring number – 1729.

Ramanujam's face lit up. No, he said, that is not a boring number at all, it is the only number that is the sum of two cubes in two different ways.

$$10^3 + 9^3 = 1729 \quad \text{and} \quad 12^3 + 1^3 = 1729$$

### *Diminishing Remainders*

Can you name the smallest number, which when divided by 10 gives a remainder of 9, when divided by 9 gives 8 as the remainder, when divided by 8 gives 7 as the remainder and so on, down to a remainder of 1 when divided by 2?

If X is the number, the problem can be set out thus:

$$X = \text{multiple of } 10 - 1$$
$$X = \text{multiple of } 9 - 1$$
$$X = \text{multiple of } 8 - 1$$
$$X = \text{multiple of } 7 - 1$$
$$X = \text{multiple of } 6 - 1$$
$$X = \text{multiple of } 5 - 1$$
$$X = \text{multiple of } 4 - 1$$
$$X = \text{multiple of } 3 - 1$$
$$X = \text{multiple of } 2 - 1$$

Therefore

$$X + 1 = \text{multiple of } 10$$
$$X + 1 = \text{multiple of } 9$$
$$X + 1 = \text{multiple of } 8$$
$$X + 1 = \text{multiple of } 7$$
$$X + 1 = \text{multiple of } 6$$
$$X + 1 = \text{multiple of } 5$$
$$X + 1 = \text{multiple of } 4$$
$$X + 1 = \text{multiple of } 3$$
$$X + 1 = \text{multiple of } 2$$

Now we find the LCM of the ten numbers. That gives

$$2^3 \times 3^2 \times 5 \times 7 = 2520$$

Therefore the required number is:

$$2520 - 1 = 2519$$

# *CONCLUSION*

Mathematical discoveries, like all others, may come from the need to solve practical problems. But a more important motive is the pure delight some individuals take in invention and discovery.

A mathematician tries to extend his powers. He knows a trick which allows him to solve one sort of problem – but might there not be other sorts that can be solved in the same way? He tries to find out how his tricks work, to group problems in families, to classify them. When a new problem comes up he will often be able to see immediately what family it belongs to, and what methods are likely to solve it.

In this book I have tried to arouse this active state of mind, to encourage mental adventurousness.

If my readers have come to share this feeling for the subject my objective has been attained.